Development Poverty and Politics

Routledge Studies in Development and Society

Development Poverty and Politics

Putting Communities in the Driver's Seat

**Richard Martin and
Ashna Mathema**

Routledge
Taylor & Francis Group
New York   London

First published 2010
by Routledge
711 Third Avenue, New York, NY 10017

Simultaneously published in the UK
by Routledge
2 Park Square, Milton Park, Abingdon, Oxfordshire OX14 4RN

Routledge is an imprint of the Taylor & Francis Group, an informa b

First issued in paperback 2011

© 2010 Taylor & Francis

Typeset in Sabon by IBT Global.

Library of Congress Cataloging-in-Publication Data
Martin, Richard, 1939–
Development, poverty, and politics : putting communities in the drive
    by Richard Martin and Ashna Mathema.
      p. cm. — (Routledge studies in development and society ; 23)
    Includes bibliographical references and index.
    1. Community development, Urban—Developing countries. 2. E
development—Citizen participation   3. Urban poor—Developing co
I. Mathema, Ashna. II. Title.
    HN981.C6M375 2009
    307.3'4416091724—dc22
    2009026448

ISBN13: 978-0-415-99562-7 (hbk)
ISBN13: 978-0-415-80797-5 (pbk)
ISBN13: 978-0-203-86208-7 (ebk)

*To Nicola, for her bottomless curiosity
and boundless support*
RM

To Sheila Resnick, with fond remembrance
AM

Contents

Boxes

Figures

Foreword

John F. C. Turner

Anyone with personal experience of impoverished environments, with professional experience of working on improvements, or with the politics of development should read this book.

The authors' pictures and portraits show and tell what life in African cities is really like for many, if not most, people today. Recently settled urban areas in other continents vary in degrees of material poverty and, perhaps, in degrees of suffering and happiness—but the underlying reality so well illustrated by the many personal stories in this book is universal. Chapters 2 and 3 describing and illustrating where and how the "other half" (*sic*) lives is a rare and rich source for thought on the meaning of development and for realistic assessments of situations and actions required for environmental improvement. Those two chapters provide yet more evidence that *the meaning of housing is in what it does for people, not just in its physical form.* Chapters 4 and 5 make it impossible for any attentive reader to confuse the meanings of housing *by* and housing *for* people: of housing seen as a life-fulfilling process, entirely different from developers' and politicians' focus on standardized commodities that are unaffordable for low-income countries, let alone for the majority of their citizens. For this reader, the next six chapters are the most stimulating of all: with reference to their own experience the authors update the old discussion of "participation"; they illustrate ways and means by which people actually can and do share decision making and responsibility for their own dwellings and environmental improvements, ending with a summary of the new management structures demanded.

This book is about paradigm change in the sphere of the built environment. Informed by their own experience, the authors describe the predominant barriers to genuine community-based development and how they can be and are sometimes overcome, endorsing as Tony Gibson's conclusion that "Local Knowledge plus Public and Professional Back-up equal Towns and Neighbourhoods that Work."[1] Patrick Geddes (1856–1932), who taught Lewis Mumford (1895–1990), understood and promoted this simple truth a century ago, but it has taken half a century since before it began to be recognized by significant numbers of forward-looking professionals and policymakers.

Aware of the increasingly critical necessity of accelerating participatory development in all contexts, the authors emphasize the indivisibility of the social, economic, and environmental dimensions that academic and bureaucratic departmentalization ignores. The authors' concern with the underlying mind-sets maintained by development-speak is clear from the number of words they frame in inverted commas—such as "slums" and "informality"—indicating that these are not the words of their choice. Wisely limited to the sphere of the built environment, this book raises the wider and deeper issues that underlie all spheres and fields of human activity, beginning and ending with the relationship between human societies and their works and nature. Whatever the nature of the sphere and field of our own work, we cannot share our experience and understanding, let alone cooperate effectively, without shared meanings of essential words. I cannot think of anything more important than getting rid of those inverted commas. Tearing down the Tower of Babel is a big job and I look forward to dismantling at least a small part of it with Richard Martin and Ashna Mathema.

Acknowledgments

No one works alone, and this work is the tip of an iceberg. Above the water, this is a limited contribution to the field of working in informal settlements. Below the water lies a massive quantity of experience, advice, and support which has molded our lives and understanding. Our mentors over many years have helped us to understand matters better through their wisdom and understanding.

For Richard, they were Bard McAllister, Steve Mulenga, Harrington Jere, and Jack Hjelt, who brought totally new insights and showed him the fundamentals of working with people; and the hugely enriching collaboration with Paul Andrew, whose enthusiasm, determination, vision, and skills made such a difference.

For Ashna, they were Anna Hardman, who got her started on this track at MIT; Jerry Erbach and Giles Clarke, who mentored her and guided her on the first professional field assignment; Bob Buckley and John Wasielewski for their advice and their confidence in her through many challenging years of consulting practice; Kiran for his unconditional and unwavering support, always; her Mum for being the magical long-distance health and spiritual advisor, and Dad, for making her a fighter, and to whom this book will be, as he puts it, "another feather in the cap."

Finally, and most importantly, we owe boundless gratitude to the people living in informal settlements—many of whom go unmentioned—who spared precious time to share their experiences; enriched our lives and work with their good humor, hospitality, fortitude, energy, and inventiveness; and inspired the fundamental message of this book. Without them, this work would not have been possible.

Thank you.

Richard Martin
Ashna Mathema

Introduction

It is unlikely that the reader will find many new ideas in this book. It is equally unlikely that everyone will agree with it. But the fact is that the development community has been looking for greater aid effectiveness for a long time, and as we wring our hands in exasperation about the large number of poor people in the world and the conditions in which they live, there is an increasing call for answers.

Why does so much aid result in so little? Why is there so much confrontation on the subject of slums, of urbanization, and of the environmental and economic impact of cities? Why do so many good people work so hard to better the lives of the poor and disadvantaged, yet the problem not only remains but in many places has intensified?

This book is an attempt to answer some of the questions and to bring together the creativity of the thousands of minds who have applied themselves to the problem of how to make aid work better. These include academics, the staff of development institutions, practitioners, and governments. But most of all, *it includes the poor themselves*. The answers are there, and the ideas have been tried and tested but remain scattered among a multitude of conference papers, journal articles, office reports, and perhaps, most depressingly, in successful projects that remain isolated enclaves which are not replicated. What is missing is the glue and the rationale to bring the diverse but consistent pattern of evidence on the subject of how to make things work.

In the process of doing so, we will be compelled to make judgments about what works and what doesn't work. Inevitably, well-respected (and equally vilified) development institutions will find some of their models, practices, and prescriptions questioned.

FOCUS OF THE BOOK

The title of the book is exceedingly broad in some sense: *Development* itself is a charged word, and to anyone inside or outside the development circles, the term can mean many different things. We will come back to this a little later, but for now, we begin by drawing a line around our ambitions to equip the reader with a better sense of our focus.

First, this book is not a theory of development economics, and it does not purport to provide the answers to a host of major issues which are troubling the development aid community such as illiteracy, the threat and impact of HIV/AIDS, the environmental consequences of global warming, and so on. Instead it focuses on one single, but very important, aspect of development, namely *urban development and the needs of the urban poor.*

This has value in a number of ways, in that it integrates physical, human, and economic development and ultimately has a major impact on all sections of society. There are also lessons from this which may be applied to rural development, the world of microfinance, the question of the management of health services, and so on. Certainly, we do not hesitate to borrow from other sectors when the experiences seem to be apposite. However, by focusing on the sector that we know best, and one which has had the advantage of decades of trial and error, we can make the most useful contribution.

Second, although this book concentrates on urban development, our discussion inevitably strays into other fields—for example, governance, community participation, construction management, housing design, and so on. These are essential components of an effective and all-inclusive system: one which passes our test for success. We therefore touch these fields to interrogate the impact that such matters have on the whole system—for example, the degree to which house design and housing standards affect the interests of the users, and how. Hopefully some of the answers we obtain from such enquiries will surprise the reader and will stimulate further thought and debate.

Third, and most dangerously, the book goes into the field of human behavior, as this is at the core of all development. If development aid doesn't work *with* people it cannot succeed. But it is in trying to understand what people respond to, and how to make things work better, that most mistakes are made. It is taken for granted that people will always welcome aid, as it is "something for nothing" and "good for them," but all too often things turn out differently. This we ascribe to imposed systems which are constrained by institutional limitations and inappropriate rules. The book is therefore, in many ways, *a search for methodologies which reflect the values of the people who will be the participants in and beneficiaries of the development.*

Finally, the discussions and debates posed in this book are based on our experiences. While several Asian countries have provided a backdrop to our work, the majority of our work has been in African cities; as a result, African issues inevitably feature strongly in the book. Discussions with colleagues who have specialized in other parts of the world reveal a common thread in most development problems, but woven into that thread are the different colors and textures of each country and culture. So while we make no secret of using African examples from time to time, we invite the reader to distil the essence of the thought and apply it to other situations with which they may be familiar.

That said, it is unnecessary to make any apology for using Africa as our focal point under the current circumstances. After all, Africa has the world's highest urbanization rate and sub-Saharan Africa has about 190 million people living in slums, the highest proportion in the world.[1] Furthermore, 70 percent of the people defined by Paul Collier as the "bottom billion," who are "falling behind," live in Africa. Africa therefore constitutes the core of the development challenge.[2]

POLICY AND PRACTICE: FASHIONS, FADS, BUZZWORDS, AND BEST PRACTICES

We argue that development principles seem to follow fashions, and, just like *Vogue*, by urging the adoption of the new season's trends, the fashion leaders imply that previous fashions should now be discarded. The housing sector presents a fascinating example of the changing trends among donors, with their support for state-subsidized sites-and-service projects in the 1970s, to slum upgrading and cost recovery in the 1980s and 1990s that emphasized the role of government as the "enabler, not provider" of housing, to very little or nothing being done in housing in the early 2000s, to now, when all of the foregoing are being questioned and revisited and new solutions are being devised. Through our experiences, we illustrate the missing link between policy and practice, between what people want and need versus what planners decide is "good" for them. This is a very fundamental problem in development that might be exacerbating the problems associated with "slums" and "informality" rather than addressing them. In this book, therefore, we try to tease out some of the threads of success that have survived the fads and fashions and can be made the foundation for a more solid and less ephemeral approach to development.

There are several tests for success in development practice, and it is—disturbingly—these very tests which have, in some cases, driven the fashions. For as the fashion for objectively verifiable results as demonstrated by quantifiable outputs requires one management system, a system that measures success by consumer satisfaction indices will require another. A system which is designed as an auditing tool to limit or prevent corruption may generate a development system very different from that which is designed to improve the welfare of the beneficiaries.

But whatever tests for success are used, we feel it essential that we should clearly state what our criterion for success is—development should result in an improved environment (in the broadest sense) for the poor. Some aspects of such improvements can be measured, and this is to be welcomed. But others, possibly equally important ones, may not be measurable; or if measurable, not measurable without complex and expensive surveys. For example, if the end result of the development is that a community has the self-confidence to better manage its own affairs, how do you measure such confidence? Can we say, with a straight face, that such changes are not beneficial, and might be even more important than some improvements to the physical—and easily measured—conditions in which people are living?

Readers of this book will encounter many familiar words. They may even, with a sigh, say to themselves—it's all the familiar material on . . . whatever. That is why we started the book by saying that very little in this book is new. But while we acknowledge this at the outset, we equally ask that the reader look afresh at the words and concepts that we use, because we think that far too often words and methods have been used without a full understanding of their significance and true meaning. The fashions that we referred to earlier must take some of the blame for this, and as people and agencies scramble to fit their ideas into the latest fashion, the words that they use and the practices which they follow may be devalued.

DISENTANGLING "DEVELOPMENT"

So, what is "development"? The corridors of academia, the coffee corner in our offices, and sometimes the corridors of power echo with debates around the nature of development. It used to be common to refer to "underdeveloped countries." So what is underdevelopment? Is it that the trains run late, that they don't run at all, or that there are no railroads? Is it that the governments are too weak or too strong?

And what does development involve? Is it something that one party does to another—"we are going to develop a new port"? Or "we are going to develop that rural area by building roads and markets"?

For some people, for example, the residents of a smart suburb, development can be a threat to their investments. To others it can mean jobs and prosperity, as when a new factory is developed in an area of unemployment.

Is development something we do to or for someone else, or can we develop ourselves? And does development involve transforming a system or economy from one state to another? Many would certainly think so, and a retrospective study of the structural adjustment programs of the 1990s would claim that this is not always for the best.

Another question of great importance is who is developing and for whom? The following quotation provides a useful perspective on the matter:

> I thought about these and many other aspects of the problem [of advancing the welfare of Africans in Uganda], feeling more convinced than ever that it is impossible to introduce to other people in other circumstances those things in life that one considers desirable; such other people do better to be left alone to find out for themselves what suits them. Then they adopt what they know and want, to what they

gradually discover by trial and error that they do want in their new setting. Little is achieved ... by those who 'have' being softhearted and solicitous towards those who 'have not'. The best things in life are learned the hard way. And need there be unanimity about what everyone means by 'best'?[3]

There is even a more extreme view that external aid is, by definition, a gift, and distorts both the values and the economy. By this very fact, aid has unintended and less than beneficial implications.

When resources are received from abroad for nothing the valuable process of generating them is lost. When resources are both generated and used locally the personal qualities and attitudes, social institutions and economic opportunities required for their employment are encouraged to develop simultaneously. The interaction of these elements of social constituents and processes then serves as a basis for further material progress.[4]

Implicit in much use of the term *development* is the concept of a norm. Either you are developed or underdeveloped. What are the indicators of this norm? Is it income, governance, the standard of services and infrastructure, or are social norms the ones that matter? Is a country that treats its old people well, or has less crime, more developed than one which does not? Once again this is a topic that is perhaps as old as time and with perhaps as many alternative approaches as one can imagine, from the Gross National Happiness of the Bhutanese to the more quantitative but still ethereal Human Development Index of UNDP.

There can be little doubt that there is good and bad development, and there are highly developed and less-well-developed countries. This much can be asserted without fear of contradiction. But beyond that, the concepts and value judgments around development are open to so many interpretations that consensus will never be achieved.

There are strongly held views about the economics of development—does wealth trickle down, or is the bottom-up system more effective in poverty reduction? Is there agreement on how to stimulate economies in such a way as to reduce poverty? What part do the politics of what is labeled the right wing and the left play in development theory? Does one look after the interests of the poor and the other those of the rich? If so, what are the politics of those who want to look after the interests of all people?

To return to the issue of good and bad development, how can these concepts be defined? What does bad development consist of? Is it wasteful, does it have no economic impact, does it harm the interests of some people? Some sorts of bad development are easy to identify:

A development which spoils ten square miles of countryside will be the work of a few people neither particularly sinful nor malevolent. They may be called Derek or Malcolm, Hubert or Shigeru, they may love golf and animals, and yet, in a few weeks, they can put in motion plans which will substantially ruin a landscape for 300 years or more.[5]

As for good development, what is it? Is there some output which characterizes it? Is it distinguished by the way it's designed and processed? Are the people affected by the development better-off after it has taken place than before? Is development a process or a product? Would Derek's scheme have been built if proper process had been followed? What is proper process and how do we ensure that it is followed?

The book does not try to answer these questions directly. Any such generalizations can always be questioned and there will never be enough common ground to overcome the demands of the specifics of local circumstances. Nevertheless, we think there is sufficient empirical evidence to construct some principles on which good development can be based. How these are applied must be left to practitioners. But, as the title of this book suggests, we believe that development poverty and politics are intertwined; and poverty cannot be addressed effectively unless the style of development and the politics of development are appropriate.

The rhetorical questions just mentioned are intended to highlight the values implicit in concepts of development. Clean tap water, universal education, and a steady job are all objectives which no one can quarrel with. But development can never be about the attainment of a universal package of uniform goals—although the Millennium Development Goals might suggest otherwise—as such goals will never, in themselves, be either universally achievable or agreed by everyone as being the most important ones.

Amartya Sen[6] overcomes this difficulty by characterizing the goal of development as the achievement of the freedoms necessary to live a life more fully. A reduction of the strain of everyday chores can result in real benefits in terms of personal health and well-being and in economic opportunity and growth. For example a woman who must spend an hour a day collecting water from a distant point will be freed from such time-wasting activities if she has her own tap in the backyard. Moreover, if the water source she used to use was polluted, she saves the additional time and expense of boiling it and/or the time and expense of treating sickness in her children caused by the pollution.

It is easy to romanticize hardship, and development cannot be equated with happiness: a glance around the streets of New York or London, or even worse, along the faces of the commuters in a subway train will show that neither development nor personal income guarantees happiness. But the freedom to choose one's own path, and the freedom to escape the relentless drudgery of poverty, is surely of extreme importance. Where people's lives are clouded by a constant fear of hunger or drought, by the pain of losing children to a sickness long since conquered in other countries, or by a life in which mere survival requires a person to work sixteen hours a day, they are prisoners of poverty. To release them from this prison is true development. They will not become petit bourgeois overnight—that is obvious—but development can bring relief from the dreadful hardships and indignities that prevent them from even enjoying life and the company of their children.

Sen puts it this way:

> The perspective of freedom need not be merely procedural (though processes do matter, inter alia, in assessing what is going on). The basic concern, I have argued, is with our capability to lead the kind of lives we have reason to value. This approach can give a very different view of development from the usual concentration on GDP or technical progress or industrialization, all of which have contingent and conditional importance without being the defining characteristics of development.[7]

He distinguishes between the "opportunity aspect" and "process aspect" of freedom.

> Unfreedom can arise either through inadequate processes (such as the violation of voting privileges or other political or civil rights) or through inadequate opportunities that some people have for achieving what they minimally would like to achieve (including the absence of such elementary opportunities as the capability to escape premature mortality or preventable morbidity or involuntary starvation).[8]

The importance of process cannot be exaggerated. It is because process aspects are difficult that they are often overlooked, which, in turn, is why much of this book has concentrated on process. *Through the correct processes, not only will development focus on what will make a real difference to people's lives, it will also be implemented in a sustainable way.*

CONSTRUCTIVE ENGAGEMENT AND BAREFOOT PROFESSIONALS

Unfortunate as it is, most development agencies[9] are not well equipped to deal with development processes as if people matter: they are much more comfortable with the mechanistic approach of deciding what should be done and simply doing it. A road is needed? Then get consultants to design it, get contractors to build it, and that is that. The question of who decides whether a road is more important than water, where the road should be located, and what standard it should be built to—these are all questions that all too often go unasked.

Any analysis of policy and procedure typically starts with the role of the state because it is the only actor which has the power to make and implement policy. But the state's role cannot be defined until the needs of the beneficiaries have been addressed. Two essential aspects of their role must be defined.

The first is the beneficiaries' *effective demands*.

These cannot be determined without their participation, and to be relevant, have to be determined in terms of effective demand in relation to present and future needs. Effective demand being what they want and are able to pay for within the fiscal system to which they are subject, is something which will change over time. Thus, one year it may be for improvements to the water system, and once that need is satisfied, the next year it may be for better roads, and so on. Accordingly, such needs cannot be satisfied by a once-off survey, or even a fully participatory project-design process. The system must work in such a way that they are able to upgrade their environment and their housing on a progressive basis, no matter how long this might take.

The second is their *capacity and capability*.

There are certain things that the state is more effective and efficient at doing than unskilled communities, especially in terms of infrastructure. This is not to say that communities are incapable of making huge efforts in such matters. We have seen wonderful water schemes, including solar-powered water pumps; roads and storm-water drainage schemes; planning and plot pegging; solid-waste disposal schemes, and so on. Communities can achieve a tremendous amount using their own skills and resources. But, however remarkable these schemes are in terms of energy and determination, they are poor cousins of the real thing. Our objective should be to help communities get access to the services they need using the comparable quality of design and construction to which other citizens are entitled.

What communities therefore ask for is help in accessing these services, and a defining role in deciding what should be provided, how and where. They do not want to have to be burdened with responsibility for things they cannot do effectively: they want the state to use its resources to help them do it. Thus, they, as the clients, decide what is wanted, and others need to help them get it. Perhaps the most vivid example of the success of such a conceptual approach writ large is the way the donor consortium proceeded in their work with two of the world's most successful community organizations—the Bangladesh Rural Advancement Commission, known as BRAC, and Grameen Bank. Fifteen years ago because of their frustration with aid effectiveness in Bangladesh, the donor consortium chose to go directly to the poor directly through these community organizations rather than attempting to go through the usual government channels. Fifteen years later, these institutions now serve more of the country's poor than does the government. BRAC is now the world's largest NGO and is viewed as having been highly successful; and Grameen's efforts, which were initiated by Nobel Laureate Muhammad Yunus, are replicated worldwide. In short, models exist and their performance, when given a chance, has been nothing short of exemplary.

CROSSING THE GREAT DIVIDE

While we plant ourselves firmly in the camp of those who believe in bottom-up development, we would be the first to point out that there must be structures within which it must take place. Bottom-up development is not the same as *laissez-faire*, and requires management in just the same way as all other development does. The perceptions of the poor regarding their environment and their future are incredibly important, as are their aspirations, as Arjun Appadurai called it, their "Capacity to Aspire."[10] Still, we do not believe that it is either desirable or practical to absolve the formal structures of government and civil society of any duty of care and/or necessity to play a role.

Much of what we write about is therefore focused on the interface between the poor and the structures of government and society in all their forms, because we feel that it is in this relationship that so much of conventional development has been weak. This is not an easy subject, and while universities may run courses on human behavior, on governance, on management and a host of other relevant topics, they do not prepare people for the reality of the interface between community and state (in whatever form it may be) and its implications for the rule of law, project, or financial management. Similarly, the formal structures of the state, whether national, regional, or local government, are typically very uneasy about this interface (with a few notable exceptions). Even architects, who purport to be experts at responding to the needs of their clients, have difficulties in true bottom-up development.

On the one hand, we believe that it is, in many cases, a mixture of ignorance and fear or inertia that prevents agencies from adopting a more people-centered approach. In this they are undoubtedly assisted by accountants who will warn of the perils and cost of public participation, and the traditional design professionals whose motivations are the maintenance of high standards (or—a more cynical point of view—protecting their interests by creating exclusivity).

On the other hand, however, we believe that while this people-centered approach has been articulated for many years by scholars such as Jane Jacobs and John Turner, it is only in the past fifteen years or so that local governance and community involvement were able to play such vital roles in development. Prior to that it was common for many multilateral and bilateral projects to be implemented by project implementation units which were usually manned by foreign consultants and domestic experts who played subordinate roles. Hence, if we can solve the problems of ignorance and fear, there may well be a real opportunity to move this people-centered approach forward.

There are many stories about people power which can be used either to support or attack the involvement of communities in development. Our aim is to distinguish sense from nonsense in this matter, and while not trying to provide an analysis of good or bad management we would like to think that we make the subject less mysterious, more accessible, and more practical. Indeed, acknowledgment of what one knows and does not know is one of the first steps in learning, and this book tries to help the reader plan and manage development in such a way that it works. The book is designed to provide guidance regarding what special skills may be required and how to use them effectively.

To return to our opening point, it is tempting to say that we have found the answers. Unfortunately, that is not true. One thing which we have learned from experience is that no one has the answers, and it is one of the sadder facets of development history that far too many people have claimed to have the answers and thereby have sparked worldwide fashions.

NEW WAYS OF WORKING

In this book, we discuss new ways of working. Of course they are *not* new, in themselves. From John F. C. Turner's pioneering work on housing,[11] which opened our eyes to the potential for ordinary people to develop their own solutions and control development processes, to the multiple academic discourses on participation, the field seems well covered. But the fact remains that much of the so-called development fails to meet its underlying objectives, or the expectations of the beneficiaries. The chief flaw in so much development work has been that good ideas and appropriate systems do not have time to take root before they, the new babies of the development world, are thrown out with the bathwater of the succeeding fashion.

This book is not simply about managing projects better, or spending money more effectively, or developing more appropriate policies. It is about how we can help the poor, and specifically the residents of urban informal settlements and other marginalized groups. With that as a starting point, we concentrate on the *hows* rather than the *whats*. By this we mean that there seem to be certain threads which run through successful projects which are derived more from how the project was designed than the solution finally adopted. Thus, the specific solution is less important than the design process—and by design we are not talking about the form of physical solutions but the relationships and duties of all the parties involved in any development activity.

The *hows* concern the manner in which a program is designed, a key part of which is a knowledge of the essential ingredients. We try to unpack those essential ingredients and to show how they relate to each other. But the *what*—how those ingredients will be mixed in the final solution—is something which can never be standardized. It must respond to local conditions, whether economic, legal, human, or environmental. Taking all factors into account, every intervention is bound to be, and probably should be, different. This is not to say that there are not common factors tying most successful projects together; and it is these common factors that we seek to present.

Finally, if there is one single theme in this book, it is that development is *not* a complicated matter, and that more often than not it can run itself provided that the relationships between the role players are structured in such a way that the interests of the ordinary beneficiary are paramount. In other words, Turner is right: housing is not a noun; it is a verb. It is a word that characterizes the importance of putting the beneficiaries, rather than a passing development fad, in the driver's seat and helping them, not dictating to them, to achieve their goals.

Part I

1 Righteous Indignation
The War on Poverty

Much of the response of the world to the hardships of life in informal settlements has had the unintended consequence of marginalizing them further by the use of labels which have pejorative connotations. These labels are used to denigrate the settlements and justify repressive action.

There are millions of people living in so-called slums. What response does the word *slum* evoke? For many people there is revulsion, fear, and occasionally outrage. To this we could add an image of filth, crime, disease, slovenliness, helplessness, and poverty. The "war on poverty," a phrase commonly used by world leaders and development practitioners, thus often translates into a war on the poor, sometimes intentionally and at other times unintentionally.

In this chapter, we look at the roots of the stigma associated with the word *slum* and the effect of policies of slum clearance and urban renewal. Subsequent chapters will show that first impressions can be very misleading, and that though the residents of so-called slums face many hardships, "slums" can be places of hope and opportunity. We also illustrate the variety and depth of personal circumstances of the residents of slums.

A HISTORICAL PERSPECTIVE OF SLUMS

Those of us who call ourselves urban planners can trace our professional past back to slums. It was the unspeakable conditions of the slums in the nineteenth century that gave rise to the early planning legislation and a realization that urbanization, the markets, and local government had to work within a new framework for the public good.

This was a necessary and good thing, as it led to the development of the principle that individual rights must be limited in the interest of the health and welfare of society as a whole. At even the lowest level—the reduction of the spread of disease—this was self-evidently of value, but as time elapsed, the coverage of legislation began to address the welfare of the individual as well.

Thus it is that the urban planning profession traces its roots back to the days of the industrial revolution, when a combination of rapid urbanization, excess demand, exploitative employers and landlords, and ignorance regarding the sources of disease turned the urban areas into places of unspeakable squalor and misery. The response took time to develop. Debates concerning the right of governments to intervene were balanced against the patently unsatisfactory living conditions being faced by the poor. Marx certainly had right on his side when he pointed to these conditions as a product, and an unacceptable product at that, of capitalism.

Eventually, the first Public Health Act was adopted in Britain in 1875, which for the first time gave the state the right and duty to interfere in the private rights of the individual (in this case, the landlord) in regard to how he was to develop his property.[1] Even to this day the line between the duties of the state and the individual are subject to similar debate. As we shall see following, the definition of the respective roles of government and private interests has a major influence on how cities develop and how residents within them view authority. But that discussion belongs to a later stage in this book: here we need only point out that it was at that time that the word *slum* was first used to describe places where unsanitary conditions, human suffering, and deprivation were the norm.

In those early days, the suffering took many forms: overcrowding, a lack of sanitation, smell, infestation by vermin, badly lit rooms, poor or no roads, and so on. To those for whom slums were home, the living conditions were accepted as a necessary evil. In those days the links between health and the physical environment were poorly understood, so slums, together with the exploitative conditions in the factories of the industrial revolution, became a recognized component of urban living. Outsiders could only react to them with revulsion. This revulsion brought with it a certain frisson of horror, which explains the success of one of the books which first brought the conditions in slums into the public eye: Jacob Riis's famous polemic about the slums of New York illustrated with his own photographs, entitled *How the Other Half Lives*. His description imparts some of the truly awful conditions at that time. The following is an extract from a report to the legislature of 1857:

> Large rooms were partitioned into several smaller ones, without regard to light or ventilation, the rate of rent being lower in proportion to space or height from the street; and they soon became filled from cellar to garret with a class of tenantry living from hand to mouth, loose in morals, improvident in habits, degraded and squalid as beggary itself . . . Rents were fixed high enough to cover damage and abuse from this class, from whom nothing was expected, and the most was made of them while they lasted. Neatness, order, cleanliness, were never dreamed of in connection with the tenant-house system, as it spread its localities from year to year; while reckless slovenliness, discontent, privation, and ignorance were left to work out their inevitable results until the entire premises reached the level of tenant-house dilapidation, containing, but sheltering not, the miserable hordes that crowded beneath the moldering, water-rotted roofs or burrowed among the rats of clammy cellars.[2]

Box 1.1   The slums of New York—1890.

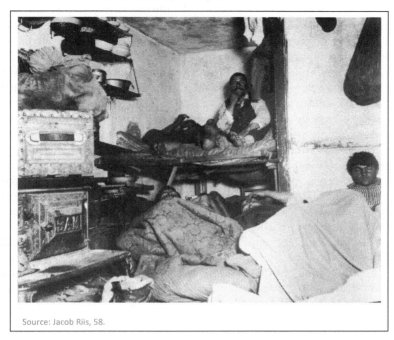

Source: Jacob Riis, 58.

The struggles of the inhabitants of the nineteenth-century slums provided good material for novelists, as mothers coped with hardships of their daily lives, not least of which was the ever-present and thriving horror of tuberculosis, dysentery, lice and rats. Slums were also associated with prostitution, drunkenness, and, of course, crime, of which the pickpockets described so graphically in Dickens's Oliver Twist were a comparatively minor problem. Not only were the residents very poor; they also had little to lose, so crime flourished. The horrible housing conditions conspired with a generally disaffected population to thwart the limited efforts of a then just emerging police force to catch the criminals.

The response of genteel members of society to the situation was, rather as it had been with slavery only fifty years before, to wring their hands in righteous indignation that people should be subjected to such horrors and then to ask—what can we do? Because, although the Public Health Acts required substantially improved standards for new dwellings, they had limited power over existing housing. Thus, while conditions were improved for many people, for very many there was little change. Even if investment in improved infrastructure such as roads and drainage made a substantial difference in time, they did not eliminate the lack of daylight, poor sanitary conditions, and overcrowding that were responsible for a substantial portion of the health problems the residents faced.

It took another fifty years before society gathered courage to contemplate the truly interventionist act, and another thirty before the political environment permitted these ideas to be put into law. We speak, of course, of the concept of compulsory purchase of the slums from their private landlords, and their subsequent demolition.[3]

The euphoria and idealism of the 1950s cannot be understated. At last there was a chance to start afresh and eliminate the shameful aberrations of nineteenth-century urbanization. Enlightened policies gave every family the right to a decent house—well-lit, dry, and big enough for every member of the family to have privacy—located in a decent environment where there was room for children to play, and there were schools, clinics, and jobs to meet everyone's needs. Thus, in Britain, the New Towns, which had been a gleam in the eye of many planners since the turn of the century, at last received the political support and funding that they needed. The destruction of so much housing by bombing in the Second World War acted as an additional spur to action.

In this way society could cleanse the inner city of the so-called slums, send many of the poor and disadvantaged to the New Towns, and, in due course, build modern and clean-looking tower blocks to rehouse those who would remain behind. Thus came about slum clearance and urban renewal, which were considered wholly "good" things.

It was not long, however, before sociologists and social workers began to notice that something was wrong. Whereas the communities in the slums had been mutually supportive and provided a safety net for a whole host of circumstances, those in the new towns and the tower blocks were alienated and confused. Gone were the days when a granny could look after the children; when the neighbor would gladly lend sugar, or even money, to help out; when the children could play in the streets in safety because everyone was looking out for them. Instead, families found themselves surrounded by people, but lonely; those with children were in a state of anxiety because the community no longer had a role in guarding them as they played. The lack of community spirit and any extended family system began to affect older children too, who became alienated from their parents, and took pleasure in acts of defiance and vandalism.

There will be many who will disagree with this account of the brave new world of housing which Britain created in the period 1950–80. But many will recognize within it similarities with the situation in many developing countries today. In most urban areas in the world there are areas of deprivation, quite easily identified by their physical characteristics. Their most common feature is that they are "informal"[4]—that is, built without the approval of the urban authorities, and not in compliance with the "formal" requirements of the law.

A BLEAK PORTRAYAL OF THE "OTHER HALF"

Because informal settlements look different, many would say they are eyesores. What is more, because the residents are perceived to be the poorest members of society, and because living conditions are bad, they are typically labeled problem areas. If there is crime in the city, then the criminals are assumed to come from those areas. If there is disease, then the bad conditions in these areas are responsible for providing the conditions in which it can spread.

If we are to characterize the relationship between established authority and informal settlements, it is in terms of "Us" and "Them"—a relationship beautifully characterized by the title of the book to which we referred earlier—*How the Other Half Lives*.

We follow the rules, *we* live in nice strong houses, *we* have roads and street lighting. *We* have children who are well behaved and go to school. *We* have regular jobs.

The other half—*they*—live in squalid conditions. *Their* ragged children pester us for money and vandalize the place. *They* carry diseases and have lice. All the criminals live in *those* places. *Their* houses are dark and dangerous.

It is hard to grasp how powerful and self-reinforcing the "them" and "us" perceptions can be. We interpret visual messages reaching our eyes to demonstrate that the situation is bad and must be dealt with. A recent book on the power of perception, *Blink: The Power of Thinking without Thinking,*[5] makes a powerful case for the accuracy of first impressions. But there is an equally compelling message from the book that sometimes first impressions are conditioned responses which can be very misleading—for example, in a famous shooting of an innocent man, a Guinean, in the Bronx in 1999 by the police because he looked sinister.[6] In this case, the policeman who shot him was so conditioned that he trusted his first impressions and felt sure that he was in an unsafe situation in that there was a black man who appeared to be ignoring him. It is not stretching the analogy too far to state that there is much the same reaction to informal settlements by better-off members of society who have an almost visceral reaction to the apparent chaos and poor quality construction of informal settlements.

The current UN-speak about eliminating slums acts as a subliminal spur to many governments to act against informal settlements—thus inevitably fueling an antipoor sentiment. A general call for action has been made. The Millennium Development Goals (MDGs) use certain standards to define a slum. Any settlement or community that fails to meet these standards is included as part of the problem to be solved, and international support has been rallied behind the concept of "eradicating" slums.[7]

The Cities Alliance has branded one of its programs "Cities without Slums." The call for action has been echoed by all agencies involved in development and strongly endorsed by the MDGs.

Slum "eradication" is spoken of with the same ease and determination as, say, smallpox eradication. The difference, however, is that most of these so-called slums are really the only economically viable housing solution for poorer communities. And unlike smallpox, the problem does not end once the slum is "eliminated"; rather, it only scratches the surface and in many cases aggravates the problem further. But more on that later in the book. For now, let it suffice to say that this righteous indignation is usually not malicious: indeed, it can be prompted by the best of motives. However, it has the effect of highlighting slums as problems, instead of (as the coiners of the slogans would have it) emphasizing the need for assistance and support to the poor. That said, there are those who really do have bad motives, who believe that slums breed crime and spoil the beauty of their city, who have no interest in helping the residents. All that interests them is to get rid of slums.

Whatever the motivation may be for describing slums as problems, the residents of these problem areas are adversely affected by such labeling.

First, the *stigma* associated with slums. Basic facts, such as having no recognized address, can negatively impact a person's sense of self and self-worth. Such people can feel, *and are in fact* excluded from society. Living somewhere which is officially described as undesirable only makes matters worse. Hence, it is not unusual for a person who lives in such areas to pretend otherwise.

Second is its impact on *self-image* and self-worth, which is another angle that emerges from the book *Blink* concerning the degree to which people's self-perception affects their performance. People who have a sense that they are inadequate perform worse than those who do not. The book *Blink* refers to an experiment in which two groups of students sitting the test used for admission to graduate school had scored equally in one set of tests but had widely divergent scores in other ones. The reason? When the students were asked to identify their race on a pretest questionnaire, that simple act was sufficient to prime them with all the negative stereotypes associated with African Americans and academic achievement—and the number of items they got right in a multiple-choice test was cut *in half*.[8]

Third, this negative self-image, in turn, affects people's performance *as well as* their *opportunities* in the territory of the formal sector. It is well known that exclusion can have other effects. Exclusion from a group is one of the harshest sanctions that people can use against a member of a group. For example, the stigmatization of people living with AIDS is considered as a seriously exacerbating factor in the rapid decline in their health. While no one could pretend that living in a so-called slum will have the same profound impact on a person's life as living with AIDS does, we must be aware that this social exclusion does have an impact which in some people can have psychological as well as economic and social consequences.

Fourth, there are numerous examples of *discrimination* against the poor that affects their chance of getting good medical care, schooling, and a host of other social benefits to which they are entitled. It is our contention that labeling people's neighborhoods as "slums" contributes to stigma and prejudice.

Fifth, there is a lot of evidence to suggest that people behave in conformity with their perceptions of their neighborhood. So if they perceive it as substandard they will treat it as such. This, in turn, drives the *environment* further down, both from the point of view of perceptions and from the objective way in which it is treated. For example, in a loved neighborhood in which people take pride, they will not drop trash or write graffiti. The opposite is true of an area in which people take no pride or interest.

This leads to the sixth issue, *crime*. The author of *Blink* makes a very convincing demonstration of the impact of environment on behavior in *The Tipping Point: How Little Things Can Make a Big Difference*. In brief, it is that where the environment looks run-down and uncared for, crime is more likely than it is in a well-kept environment.

> The criminal—far from being someone who acts for fundamental, intrinsic reasons and who lives for his own world—is actually someone acutely sensitive to his environment, who is alert to all kinds of cues, and who is prompted to commit crimes based on his perception of the world around him.[9]

Evidence of this can be found even now in developed economies. Look at the case of Washington, DC, where three of the city's four quadrants were considered "unsafe" until very recently. These are mostly poor neighborhoods, which are ill-maintained and associated with high levels of crime.

Social exclusion is not the only barrier that the residents of slums face. Although there is evidence to suggest that this is changing thanks to international pressure, in many countries residents of informal settlements are denied *access to basic services* like water and sanitation on the grounds that they are illegal. Even if they are not actively denied access to services on principle, a huge number are in fact deprived of such services because there are no funds to provide them. Few systems of governance can withstand the power of interest groups in local politics which insist that priority should be given to those who live in authorized settlements and pay their taxes.

As we know, water is one of the most effective routes by which disease is spread, and if there is no clean water then it inevitably follows that the residents will be affected by intestinal diseases. In adults this is often controllable, but for children it can be life-threatening. Even if it is not, it affects their well-being and nutritional status. Such communities are also vulnerable, of course, to the more severe diseases such as cholera and typhoid.

If basics such as water and sanitation are not provided, it is not surprising that roads and storm-water drainage are not either. Good roads are a matter of convenience, and living in an environment without them can be extremely difficult. But storm-water drainage is far more important than that and is another factor which plays a major role in health and safety. In malarial countries, stagnant water allows the mosquitoes to breed with obvious consequences. The impact of this on health is extraordinarily huge: infant mortality rises rapidly, adults suffer periodic bouts of debilitating and unpleasant disease, and the community spends a disproportionate amount of money on health care—often of a palliative nature.

Later in this book, we tell some stories to demonstrate that the perceptions of slums as hotbeds of crime, dirt, and indigence are very misleading. On the contrary, given the right environment—in the political sense—residents of such areas build flourishing societies which work every bit as well as those in the formal areas of town, and often better, considering the limited resources they have. Suffice it to say here that the negative perceptions which are engendered by calls to do something about slums can be very damaging to the very people who are supposed to be helped. In this regard, the rhetoric on addressing the slum problem through the MDGs is dangerous in that it can easily be—and is—misconstrued by governments as a simple directive to do away with slums.

"SOLVING" THE PROBLEM

Slums, in the most basic sense, are an indication not of the failure of those who live in them but rather that of the governments in performing the basic functions they are meant to, and of society at large. It is not strange, therefore, to see their attempts to hide their shortcomings by painting a false image. For example, when the President of India was visiting an area with a rather squalid squatter settlement, the day before his arrival the authorities removed all the roadside informal businesses and simply erected a blue plastic wall along the side of the road to screen the distasteful sight. Within hours of his departure, the screen had been torn down by the squatters.[10]

This sort of government reaction to slums and shanty towns is not uncommon. In 1991, Bangkok hosted an international conference that brought over ten thousand delegates from more than 160 countries to the city. In the months before the event, the Thai government forcibly removed over two thousand slum dwellers from the areas immediately surrounding the new $90 million Queen Sirikit National Convention Center, which hosted the conference. Hundreds of shanties in informal settlements were destroyed and a huge metal wall was erected to conceal the devastation left behind. Similarly, in 1976, when Philippines hosted another international conference, President Marcos initiated a "beautification" campaign in which four hundred families were evicted from slums in Manila during the months preceding the event.

> Despite the array of slum improvement programs . . . Thailand and the Philippines both relied on shortsighted strategies of forced removal in order to conceal the existence of slum dwellers and, in doing so, protect national claims of "development."[11]

Many people feel that in the long run slum clearance is the right approach—move the people and let them occupy decent modern housing. Such calls are even louder where the settlements encroach on public facilities and services such as railway lines, port areas, parks, and school sites. And, it is said, by so doing we will improve people's health and reduce crime.

Many people feel strongly that informal settlements should be removed completely, because they are an eyesore[12] or because they occupy high value land, which should be put to uses which are more profitable. It is also argued that it is impossible to simply upgrade the settlements by the construction of roads and supply of water and sanitation, due to the complexity of the layouts of many informal settlements, or the very high densities which make the construction of any infrastructure difficult.[13] The resettlement argument has the appeal of simplicity: a clean sweep of the problem.

We shall not devote much space here to the technical issues involved in resettlement, or the phenomenon of resettlement which so often occurs where sites have been cleared. Instead we make four points. One, new housing is always more expensive (in terms of rental or purchase price) than the informal houses it replaces. Two, new housing often provides less space: typically, apartment blocks are built to rehouse the families, and even where the housing unit may be of similar size, the amount of space available to a family is much smaller because they have lost their outdoor space.[14] Three, where people are resettled to individual houses, the locations are usually remote, so traveling costs and time increase substantially. And finally, a point which we shall make at some length is the effect of the destruction of the social fabric of the community which is caused by resettlement.

The trauma and cost in both social and economic terms of such resettlement have long been understood. For example, in a slum clearance project in Lagos, Nigeria, while resettlement was taking place the remaining traders saw their profits dwindle, and in the new site there was not enough trade to support more than a few shops. For the individual householders the story was much the same: they couldn't afford to build to the standards prescribed in the new site, and the compensation they had been given was typically a small proportion of the cost of a minimum standard house in the new location.[15]

The following quotation sums up the issues in many schemes:

> We have to provide housing for them before we can ask them to vacate. When we try to do this, those very people, whom we seek to benefit, raise difficulties and are reluctant to move. This is to some extent understandable, for their lives and work have revolved near that area and to take them far away means to uproot them from their work. Also, whatever accommodation might be provided, is likely to have a higher rent, even though it might be subsidized.[16]

These are the words of Jawaharlal Nehru, the first prime minister of India, but not all politicians are as wise, and few allow themselves to think about the social consequences of such disruption. Summing up the experience in Lagos, Marris writes:

> Wherever communities are transplanted, familiar complaints arise—of rising expenses, longer journeys, family relationships disrupted, and business lost: the poorer they are, the less resiliently they can withstand these hardships. Town planners often act as if the buildings they provide would themselves restore those social and economic losses: if there is a community centre, there will be a community; if shops then trade. But new social and economic relationships obstinately refuse to form about the planned facilities. In Johannesburg, Lagos, or London, the same contrasts set off the old communities from the new: all the genial warmth and spontaneity of life flows into the over-crowded, dilapidated streets of the long-settlement neighborhoods, and the municipal estates remain barren in their suburban propriety.[17]

Marris's mention of London doubtlessly refers to his work with the residents of Bethnal Green—a so-called slum area of that city. This had demonstrated the very disruptive impact of slum clearance when they were resettled to the suburban fringes.[18] We mention this because it is important to avoid thinking of this as a "third world" problem.

Paradoxically, there are also forces which resist the idea of improving slum conditions. Landlords may resist on the ground that their profits might be reduced, leaders may object on the basis that their power base could be undermined, and even local elected officials might prefer the status quo on the grounds that "so long as you keep them poor and needy and dependent, you get the votes." This angle is also often responsible for the kinds of distorted programs pursued by government or other special-interest groups.

Slum dwellers, on their part, can use this righteous indignation to their own benefit. There is nothing wrong with the residents of disadvantaged areas complaining about their conditions: indeed, they have a right to do so, and there are many tools in the protest lexicon to be used in such cases. Some techniques get change, some get fame for the leader, and some get nothing except a sense of relief and satisfaction that the authorities have been embarrassed, annoyed, or whatever. Protest can yield results, but these results can be negative ones, so the strategy must be carefully considered.

And if the residents of informal settlements choose words such as *slum* to raise awareness of the conditions in which they are living, that is their choice. They can also refer to the risk which those settlements present for the spread of disease. The threat of cholera has often been used as a trigger for swift action to improve conditions in informal settlements, just as it has been for hundreds of years.

But there is a big difference between people choosing to use the word *slum* to get political attention and, hopefully, resources, and having the word *slum* used as a label of denigration. In the latter case, it can lead to discrimination at best, and destruction at worst.

2 How the Other Half Lives
Slums and Informality

This chapter provides a series of snapshots to illustrate the variety of conditions in informal settlements. There are common threads of deprivation and bad infrastructure, but there is also an overlay of achievement and development in the face of adversity.

As we have discussed, the word *slum* masks complexity and can too easily be used as a code for any undesirable—or, dare we say, lower class— housing. It was not at all unusual in the early days of site and service projects for them to be labeled slums. Passing motorists would compare the neat anonymity of public housing projects with the untidy process of self-help housing development and proclaim "not in my backyard" (NIMBY). They did not have to justify their opposition: it was enough to simply use the label *slum*.

In practice there is a huge variety of settlements which, in terms of the UN definitions, are categorized as slums. Indeed, the variations are as great as that of urban form itself. There are the Manhattans of the slum world, and there are the Los Angeleses. There are tiny communities and there are truly huge ones. They vary in terms of form, history, physical, and social conditions. Other variables include their legal status and the degree to which they are recognized and served by local government.

We begin here by clarifying our definition of "slums," which, as Chapter 1 demonstrated, so often tends to be oversimplified in its characterization. Commonly referred to as informal settlements, squatter settlements, or shanty towns, not all slums are informal settlements, and not all informal settlements are slums. The matrix in Figure 2.1 shows one way of categorizing the broad range of informality that exists in cities worldwide, based on the level of services—both housing and infrastructure—and the legal security of tenure. If either one condition is absent, it is an informal settlement. If service access is poor, it is typically deemed a slum. So, in the matrix following, the gray and black squares constitute informal settlements, while the two squares on the right are the real slums that are lacking in either the level of basic service provision

Figure 2.1   Level of informality.

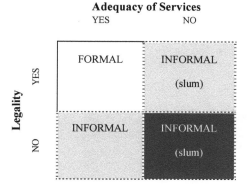

or the quality of housing itself. In other words, an informal settlement is not necessarily a slum, and a slum is not necessarily an "illegal" entity, both of which become more evident in case studies of such settlements presented in this chapter.

INFORMALITY IN CITIES

We now illustrate these various typologies of informal settlements in the context of cities, communities, and the individuals who inhabit them. Using examples of cities in different countries, we highlight how different and heterogeneous these communities actually are, and how important it is to understand their underlying dynamics in order to devise the appropriate solutions. But before proceeding, we make a note for the reader: in terms of the sample of cities presented, there is a clear tilt toward African cities, for no particular reason except that the authors have worked more in these cities.

What follows, therefore, is a series of snapshots of informality at the city level, which are then illustrated in more detail in some cases through examples of the informal settlement typology outlined previously. The idea here is to show how the underlying culture and social fabric in these places vary not just across countries and cities but also across settlements within each city.[1]

Mbabane, Swaziland

We start with Swaziland.[2] The informal settlements in the capital city of Mbabane could be assumed to be the result of urbanization outstripping the supply of formal housing, or at least formal housing opportunities. But they are, in practice, much more than that. In 2005 they represented about 70 percent of the population of the city, and effectively constitute

Box 2.1   Mbabane's informal settlements.

Source: Field research by Mathema, 2002–2005.

an alternative lifestyle. This is a lifestyle which is not hampered by petty regulation, where one can live in a semi-rural style, but still be within the city. This includes the opportunity to keep livestock, practice trades and commerce with a minimum of official interference, and construct affordable houses from traditional materials. These are not squatters: all residents have permission to occupy the land, either from the chief (chiefs being the nominal custodians of almost all land before it came under the jurisdiction of urban authorities) or the local administration. The settlements constitute a reservoir of affordable rental housing. However, they do not have access to basic services such as water, sanitation, or electricity, and mostly constitute modest houses built with local materials that do not meet the country's building standards, and are hence deemed as "slums." Box 2.1 shows pictures of different "slums" in Mbabane, where houses are made of traditional stick-and-mud walls and tin roofs.

Eritrea

Next, given a similar scale to Mbabane, we present informal settlements in cities in Eritrea. Box 2.2 illustrates some settlements which are legally occupied, some even planned, but remain underserviced with relatively poor housing quality. That said, one remarkable characteristic of Eritrea's settlements is the degree of cleanliness observed in even the poorest areas.

The first illustration is a legal settlement in Asmara, the capital city, where most plots are between 100 and 350 m². The housing quality is mixed, but in general it looks congested, with narrow access roads and

Box 2.2   Informal settlements in Eritrea.

Illustration 1   A courtyard house in Arba-ite Asmara.

Illustration 2.   A courtyard house in Dekhamhara.
Main house constructed with hollow concrete block walls, CI sheet roof, unpaved floors. Open-air cooking space, and semi-covered dry pit in courtyard. House has piped water and electricity.

View of courtyard from neighboring house   *Open air cooking space in courtyard*   *Covered dry pit, located in plot*

Illustration 3.   A courtyard house in Kutmia Settlement, Massawa.
Main house constructed with wood panels, and CI sheet roofing. Auxiliary structures made with flatted CI sheets and scrap materials. Piped water in plot, and electricity.

Cooking area in courtyard   *"Porch" attached to main*   *Rudimentary dry pit, located in plot*

Source: Ashna Mathema, *Eritrea: Housing and Urban Development Policy Study, Qualitative Study*, prepared for Ministry of Public Works, Eritrea and UN-Habitat, 2005 (Unpublished).

poor sanitation (dry-pit latrines shared by a large number of households). The second is a house in a similar settlement in Dekhamhara, another town near Asmara. The third is a house in Massawa, Eritrea's main port city, in a planned settlement where most households have legal tenure, but is characterized by poor housing quality and sanitation. The settlement has trunk infrastructure for water and electricity and wide but unpaved access roads.

Accra, Ghana

A different type of informal housing typology is that found in old Accra in Ghana. These areas were developed in the early 1900s as formally designated housing in the colonial city of Accra. Our example, Jamestown, is very well-located, in the city center, and served by streets and public spaces. Two factors, however, make these parts of Accra unique. The first is that, due to inheritance traditions, ownership of these properties has often been passed to all the descendants of the original owners (which might amount to anywhere between thirty and one hundred adults). As a result of this multiple ownership, and the fact that many of the owners are not residents, it is very difficult, if not impossible, to get consensus on selling or making any improvements to the house. The upside of this is that it has prevented downward raiding and thereby preserved excellent locations for low-income housing. The down side, however, is that upgrading efforts—in any conventional sense, or through redevelopment—are next to impossible using the conventional "community participation" model. This is because each plot effectively constitutes a community on its own, and getting consensus among owners within one plot is hard enough, let alone the entire neighborhood.

The second factor is that sanitation has somehow been sidelined in these areas. For reasons which are part cultural and part poverty-related, most of the houses in the old neighborhoods do not have toilets. The most common sanitation facility is the public toilet, operated mostly by private operators who charge exorbitant rates (see Boxes 2.3 and 2.4). Other services are also lacking, and the concentration of poverty is extremely high. So much so that in comparing these legally occupied areas of Accra to similar settlements in four other African cities, it was observed that the ones in Accra fared by far the worst.[3] That said, the point to be made here is that these settlements are clearly very different from the "slums" found in Mbabane and Eritrea.

Addis Ababa

Addis Ababa, the capital of Ethiopia, presents another interesting variation from many other cities where housing is characterized by a clear correlation between "illegality" and "informality." On one hand, many of the "legal"

Box 2.3   Evidence of lacking sanitation in Accra's settlements.

Top left: The beach front used for dumping garbage and defecating
Top right: A wall in Jamestown with graffiti warning against urinating
Bottom left: A public toilet in an informal settlement
Bottom right: An open drain used for dumping garbage and defecating

Source: Ashna Mathema, *Qualitative Study: Household Interviews*, Accra, Ghana (Unpublished research for African Union for Housing Finance, funded by Cities Alliance/ The World Bank), 2006.

settlements in the city constitute what one would term "informal" housing, and these are none other than the government- or *kebele*-owned rental
houses all across the city.

Kebele housing constitutes about a fourth of the housing stock in
Addis, and although "legal" (being public-sector subsidized rental housing), it is dilapidated, with little or no infrastructure, and characterized
by a high level of overcrowding. In other words, these are slums in the
real sense of the word, even though they are not officially termed as such.
In some sense they fare worse than slums in other cities because it is
illegal for the residents to make any improvements to the structures (see
Box 2.5). Yet, in Ethiopia, the term *informal housing* does not officially
include *kebele* housing.

A typical *kebele* house may be described as a single room, 3 to 4
meters wide and 4 to 5 meters long, accommodating between five and
ten people. The large majority is made of traditional *chika* construction—using mud and wood or straw. Sanitation and hygiene in these
settlements is poor: some households have access to communal toilets;

Box 2.4   Inner-city informality: Jamestown, Accra.

The Cost of Hygiene

In Jamestown, one of Accra's center city informal settlements, water and sanitation facilities are practically non-existent within the plots. People mostly buy water, at 2–3 times the actual cost, and use public toilets which are expensive, often in short supply, and ill-maintained.

Household expense on services (water, sanitation, electricity/ lighting) is 15–20 percent of total household expenditure and this includes zero expense for housing. Public toilets are very expensive: the price can vary between 400–1000 cedis ($0.04) per use. Expenditure on toilets is directly proportional to the number of household members, and particularly high for larger families which are also poorer. Five to ten percent of the total household expenditure is on using public toilets. In comparison, the economically better off families who have toilets in their plots typically spend in the range of 2 percent of total expenditure on sanitation. The consequence: children defecating in the beaches or the bush, or even in the open, adults using any available corner. Most importantly, there is little achievement by way of more sanitary communities when 40–50 percent of the community cannot afford or is unwilling to pay the monopolistic fees charged by the toilet operators.

Rental housing in these settlements is operated essentially by absentee landlords, who do not provide toilets or water as part of the accommodation. Despite this, it is typical for a tenant to pay 3 years' rent to the landlord in advance.

Top and middle: A view of Jamestown's main street from the second floor of a building.

Bottom: A family house in Jamestown, which has about 10 small rooms, each of which is occupied by an entire family.

The tenants lack any better or affordable alternative. Physical conditions are similar to the old family houses discussed above; in some cases, the tenants undertake basic improvements with their own funds. For the landlords, this is a low-risk, high return investment. For the tenants, it is the only affordable housing available.

Source: Ashna Mathema, *Qualitative Study: Household Interviews, Accra, Ghana* [unpublished background research for African Union for Housing Finance, funded by Cities Alliance/ The World Bank], 2006.

Box 2.5 "Formal" informality: *Kebele* housing in Addis.

Kebele houses lining the streets of an upgraded settlement

A house that collapsed due to lack of maintenance by both the landlord (*kebele*) and the user.

Source: Field research conducted by Ashna Mathema in 2004.

others defecate in the open. Household chores—grinding, drying spices, laundry—are done in the street or in the small common open spaces outside the houses. The majority of households use communal kitchens to bake the *injera* bread. Drainage and garbage-disposal systems are typically nonexistent, except in neighborhoods which have been included in past or ongoing upgrading efforts. Some settlements have benefited from NGO-funded programs that have installed communal—in some cases even individual—water points.

On the other hand, there is a large segment of informal settlements that is "illegal," but not all these illegal houses are substandard. In fact, many of these areas are well-planned if not adequately serviced. Here, the communities decided that they could not wait for land development at public expense, and that they should take the initiative themselves. They formed cooperatives, bought land from farmers, and, in full knowledge of prevailing land development standards, pegged out their plots using the government-mandated road widths, plot sizes, and so on, so that when the time came for the government to provide roads, water, and other services, they could be fully recognized as legitimate residents of the city. In many cases, these areas are well planned, only awaiting approvals from the government. But until that happens, ironically, their tenure is clearly insecure, and the risk of demolition is real. In other words, despite relatively high planning standards, they lack access to basic services and are "illegal" in that they do not fulfill minimum standards.

One example is *Kebele* 05, which, like other agro-based settlements in the suburbs of Addis Ababa, has been facing the growing pressure of urbanization and increased demand for land and housing. Formerly comprising predominantly farmers and farmland, it has become a target for "illegal" development by private cooperatives. Two settlements organized by cooperatives are presented here: the first, comprising 202 families, and another self-help cooperative that was started in 1992 with 168 households. These cooperatives differ from the formal-sector housing cooperatives in that they do not have any legal recognition. In both cases, land was purchased from farmers at minimal price, and members of the cooperatives oversaw the development of a workable site plan with basic infrastructure. Land was subdivided and "sold" to the members. Money collected from the members toward provision of infrastructure was used for the development of both physical and social services. Interestingly, the plot sizes set by some cooperatives have a clear match with the "acceptable" government-allocated plot of 165–175m² (see Box 2.6).[4] On the selling front, there are the farmers who, forced by their limited economic prospects from subsistence farming, are selling their plots to these cooperatives and to individuals. Both sides recognize that the transaction is not legal, but still choose to go ahead with it, in the hope that the government will one day legalize them. Why the settlement should be "illegal" makes for an interesting question.

Box 2.6   Illegal development or resourcefulness?

Developed by 'housing cooperatives' (unlicensed) on 'illegally' purchased land, such settlements are on the rise in Addis' peripheries. The developers have planned the layout per the building standards of the city—for example, the plot size is exactly equal to the minimum prescribed size—in the hope that one day they will get legalized.

Source: Field research by Ashna Mathema in 2004.

Dar es Salaam, Tanzania

Illegal land and housing development similar to that in Addis has taken place in Dar es Salaam, Tanzania. Here, too, traditional landowners subdivided land in formal parcels for sale to the public. These were not spontaneous invasions, but a deliberate and careful entry into the land market. The distinction between these developments and the formal sector is that no title deeds were exchanged. However, in many cases, areas within these settlements were subdivided without catering for access roads or drainage, or located in flood-prone areas. This, however, did not mean that the residents had any sense of insecurity of tenure. The government has recently embarked on a program to formalize these informal settlements: two-year (recently revised to five years) occupancy licenses are being issued as a transitional step toward formalization when residents are able to acquire a full-fledged title. Of course, this will also entail some level of planning and revisions in plot boundaries, as well as some relocation, the experience of which is still to be seen.

Dar es Salaam presents a fairly optimistic picture in terms of its informal settlements compared to many other cities. The construction quality of houses in the informal settlements is fairly good (see Box 2.7). The structures are built with mud or concrete blocks and roofed with corrugated tin sheets. Concrete blocks are typically made at home or bought from a local

Box 2.7 "Slums" in Dar es Salaam.

The top two photos show the type of houses common in informal settlements. The bottom two show the extent of flooding in some areas; stagnant water breeds mosquitoes, causing malaria.

Source: Field world by Ashna Mathema in 2006.

manufacturer within the settlement. The most common type of house is the so-called *Swahili* house: it has a central corridor, opening into two or three rooms on either side, and leads into a backyard which has the toilet/ bath and sometimes a kitchen. A small front porch is also a common feature. The porch and backyard are used for daily outdoor activities, such as washing and food preparation. The corridor is commonly used for cooking and storing water.

Most plots have pit latrines, shared among the various households living on the plot. The bath area is usually adjacent to the pit, or in the same space, such that the wastewater from the bath goes into the same pit as the latrine. Also, often this space is not covered with a roof, allowing rainwater to get in. As a result, overflows are common, resulting in seepage of human waste into the soil. There is also the common practice of using soakaways, which collect the wastewater and/or and human waste, without any septic tank. When they get full, they are "punctured," releasing the untreated waste into the ground. The close proximity of such disposal systems to wells which provide drinking water is often the cause for cholera and typhoid epidemics. As reported by one resident, these settlements are very often "the starting point of cholera and typhoid epidemics in the city." Another common problem is malaria. This is caused by stagnating pools of water/wastewater and garbage, which have no outlet due to the unplanned nature of the settlements: plots are joined to each other in a way that allows for no access roads or drainage, and no way to channel the wastewater out without encroaching on someone else's property.[5]

Lagos, Nigeria

A glimpse of the slums of Lagos presents a sharp contrast to Dar es Salaam's informal settlements. An estimated 70 percent of Lagos' population lives in slums.[6] Lagos' situation is unique, not least of all because of the massive numbers of the megacity's fifteen-million-strong population that lack formal housing and basic services, but also because many of these slum settlements are not upgradeable, being located in hazard zones. Also, the scale of disparity between the rich and the poor, the haves and the have-nots, the formal and the informal, is evident not just from the physical urban form of the city but also the fact that even middle-income households cannot afford to buy a house in the formal city.

Lagos' slums are characterized by extremely high densities, poor housing conditions, and little or no access to basic infrastructure like water or sanitation. Many of these slums are along the coastline, extending into the sea, with people either "filling up" areas (to keep the ground above sea level) or dredging others (to facilitate drainage). As a result, the ground in many of these areas is like a wet sponge, unfit for construction. There are also entire settlements built on stilts, with houses standing just above the water (see Box 2.8).

Box 2.8   Living on the edge: Lagos' slums.

As the water level has risen over the years, so has the "filling" of the ground beneath. There are areas such as Makoko, for example, that have already been filled 3-5 meters in the past 10 years, a grim reminder of people trying to "keep their heads above water," both literally and figuratively. There are entire streets where the original ground floor is now underground, i.e. what used to be the first floor is now the ground level (photo 1).

Where reclamation is not possible, there are entire neighborhoods built on stilts in water that is black and viscous (photo 2), much like the toilet scene in the movie, Slumdog Millionaire. The houses are made of wood (bamboo and/ or teak), with gaps between the planks wide enough to see the murky water below. These stilted houses are interconnected with "streets" or bridges made of precariously joined wooden planks. Small boats traverse the walkways and under the bridges, serving as transportation for the local residents and mobile vendors (e.g. women selling fresh hot foods, see photo 3).

A "toilet" is typically a wooden stilted enclosure with a hole in the floor from where the waste goes directly into the lagoon (see photo 4). This enclosure applies to those who can afford the extra wood, and the luxury of privacy. Others just defecate into the lagoon or on dry ground. It is an unbelievable sight: adults casually going about their daily business as children play around in the garbage and raw sewage. With garbage strewn everywhere, it is difficult to distinguish between a landfill and a housing development: the first comprises houses built on garbage, the other constitutes garbage lining (the streets along) the houses. And while "slums" exist in every large or small city of the developing world, it is difficult to understand how such appalling conditions can exist—and the extent of poverty and disparity—in a resource-rich country such as Nigeria. That said, when the vast majority of a city or country lives in such abhorrent conditions, it is almost always the case that the problem is a direct result of failed policy, not "poverty" per se.

A woman interviewed in one of the coastal settlements had a husband earning N30,000 ($300) per month. She lives there despite the fact that the water level rises every day during high tide, so much so that they stack their belongings up to keep them dry. Asked why she chooses to stay there in a rented wooded stilted shack with no sanitation, she said, *"There isn't anything else out there in the city. . . . if I could find something, I would take it."*

Source: Ashna Mathema, *Slums and Sprawl: The unintended consequences of well-intended regulation* (Unpublished background research for the World Bank) 2008.

Box 2.9 Real estate "not for sale" in Lagos.

It is common to find warnings like "This house is not for sale", and "Beware of 419", with which people try to stave off fraudulent sale or purchase of their homes without their knowledge.

Source: Field research conducted by Mathema in 2008.

It is estimated that more than 80 percent of Lagos' population lives in overcrowded conditions, with room occupancy ranging from three persons per room[7] to six persons per room. The density in the slums ranges from 790 to 1,240 people per hectare,[8] which is extremely high to begin with, but notably more so in light of the fact that these settlements consist of mostly one-story structures and hence have very little built-up floor space per person. The disparity in the consumption of land and resources is evident from the difference between these settlements and the rich neighborhoods. An example of the latter is Eti Osa (the Local Government Authority, which includes Victoria Island and Ikoyi), which has an average density of ten to fifteen people per hectare.

Another interesting feature of the Lagos property market is the extent of ambiguity in the title/ownership, which applies equally to formal and informal real estate. Unlike some other African cities, rather than "to-let" signs, one sees signs such as "This property is NOT for sale" (see box 2.9), to warn the gullible buyer about a potentially fraudulent sale of the property initiated by an unknown third party.

Nairobi, Kenya

Nairobi has a large variety of these so-called slums on land that is occupied both legally and illegally. In addition to true slums located on hazard zones and lacking in basic infrastructure, Nairobi has many legally occupied areas lacking in infrastructure or quality housing. We illustrate this with three settlements: *Kawangware, Gitare Marigu, and Lunga Lunga.*[9]

Kawangware is an example of a settlement in which most of the land is titled. However, the area lacks services and suffers from a high level of unemployment and poverty, which have together resulted in very poor living conditions (see Box 2.10). According to the 1999 census, the population of Kawangware was nearly 200,000. Most households in this settlement have legal titles to their plots. Many houses belong to absentee landlords who have built several rows of rooms to rent out. Rents range from KSh600 to KSh2,000 ($9 to $30), depending on the quality of the structure and availability of services.

The plots have water connections, but the supply is irregular. Water comes on one or two days per week, sometimes not at all. Most households buy water from vendors. Sanitation is extremely poor in the settlement. Most plots have their own dry-pit latrines, but each latrine is shared by ten to thirty households, depending on the number of rental units on the plots. According to a local leader, there are several cases where the plots do not have any toilets or bath facilities. The health workers monitor this and require the owners to construct the facilities, especially in cases where there are renters on the plot. The more well-to-do households, and those located on the main streets, have flush toilets connected to the city's sewer system.

Most plots have their own individual metered electricity connections. These connections are sometimes extended to the renters on the plots, for a flat fee per bulb/electric point, sometimes with restrictions on the time of usage. There are no designated areas for dumping garbage within the settlement. The local community organization occasionally organizes efforts towards environmental education, where people may get together under the guidance of the health workers to clean up the settlement—which essentially means collecting and burning garbage.

Gitare Marigu is a village in the northeast of Nairobi. It comprises three communities, one among which is Kinyago Gitare Marigu, comprising residents of Kinyago village resettled here to make way for a police station in Kinyago (this plan eventually did not materialize). Kinyago village was near a dumpsite, where migrants from villages had settled. The main source of income was scavenging from the dumpsite, providing a source of food and income from sale of recyclable goods. Under the resettlement arrangements temporary residence permits were awarded to each beneficiary household, granting them the right to build temporary shelters, until further notice—in other words, until the government decided what to do with them. That was in 1993. During the time of the site visit in 2005, the community still did not know what their status was, that is, whether they were to be moved, and if so, when. As a result, few have ventured into any substantive home improvements (see Box 2.11).

Box 2.10   Case of a "legal" slum: Kawangware, Nairobi.

Top: Typical rental buildings with rows of rooms, each of which is occu-
pied by a single family.
Bottom left: Garbage is strewn around the settlements
Bottom right: A typical pit latrine found on the plots to serve the rental units

Source: Field research by Mathema in 2005.

The residents were resettled to this site, which is designated open space
in the master plan. It is located on a hillside, which slopes down into the
Nairobi River. The layout has about eight rows of plots (average plot size
4.5 m wide and 6 m long) along the slope. Six 1-meter-wide footpaths run
at intervals—after each pair of adjacent plot rows—through the sloping
length of the site, to provide "access" to every plot. These footpaths are
extremely narrow by all standards relative to the number of houses they are
servicing. At the beginning, the new settlement did not have any provision
for water supply, sanitation, or electricity. The only source of water was a
natural spring, where people queued up each morning with their vessels to
fill water for consumption. It was common practice to defecate in the open.
Over the years, informal systems for delivery of water and electricity got
established, and people built toilets for shared or personal use.

An estimated 10 percent of the households have individual water connections. These households have set up water-vending businesses to the rest
of the settlement (@ KSh3–5/20 liters). During dry spells when water is in

short supply, the spring is used. The last dry spell occurred a month before the interview for a period of two weeks.

Less than 5 percent of the households have toilets. The waste from these toilets is discharged directly onto the ground outside the house (see right photo in Box 2.11) and thence into the river. Common/shared toilets are located in one stretch of land under the electric power lines where house construction is not permitted. There is often a problem of overfull pits and rat infestation. Households that are located far from these toilet facilities use the common practice of "flying toilets"[10] or the bush. Some have built public/shared toilets straight above the city sewage line at the edge of the settlement.

Many households draw connections from three to four legally installed metered electric connections adjacent to the settlement. There is a restriction on the time of usage, however: only between 6 pm and 9 am.

The unpaved footpaths serve as the only access during emergencies. During rains, the water from the street level comes gushing down these channels with much force. This is clearly visible from the erosion that has occurred as water has formed natural channels to get to the river (middle photo in Box 2.11). Also, some of the houses are located very close to the river and hence face risk of frequent flooding during the rains.

Lunga Lunga, on the other hand, is an illegal squatter settlement established near a dumpsite in the early 1970s (the dumpsite was later abandoned) and Nairobi's industrial area. In the early days, Lunga Lunga provided a suitable site for people scavenging in the dumpsite or looking for jobs in the factories. The original residents were farmers from the surrounding villages within Nairobi, who built temporary houses using cartons and scrap. The settlement has a population of approximately 150,000 and constitutes a wide range of people, successful entrepreneurs

Box 2.11 A government resettlement project on an illegal site: Gitare Marigu, Nairobi.

Houses in the settlement along the hillside overlooking Nairobi River	One of the walking paths, which suffers major erosion during rains	A pipe from a toilet, disposing waste into the open

Source: Field research by Mathema in 2005.

Box 2.12   An illegal informal settlement: Lunga Lunga, Nairobi.

A street with an open drain	A typical street with linear blocks of rental housing made of tin sheets	A public pay-to-use toilet/ shower facility, owned by a local resident

Source: Field research by Ashna Mathema in 2005.

living alongside jobless individuals. It is unplanned and congested for most part, and lacking in basic infrastructure. (see Box 2.12).

Over the years, as demand for housing has grown, people in Lunga Lunga have recognized the income-generating potential of rental housing. This led to a wave of new construction in 1986–87. Having established de facto rights over the land they occupied, many original residents also began to "sell" the land (partial or whole) to newcomers and outsiders. The transaction was based on the premise that the "buyer was paying the seller for the materials invested on the property, not the land." At the time, a house with eight rooms, constructed with tin sheets, could fetch as much as KSh150,000 ($2,150). According to the leader of Lunga Lunga, from 1991 to date, this settlement has been among the fastest growing areas in Nairobi. This is due to its close proximity to the industrial area, and the fact that transport is quite expensive in the city.

Water sale is a big business in Lunga Lunga. Of the estimated 30,000 to 40,000 households, only ten have water meters. To get greater coverage, they have created additional outlets by extending the pipes to other parts of the settlement. The sale price of water is KSh2 per jerry can (20 liters). According to the community leader,

> It costs some KSh3,000 to get a water meter. If you have money, you apply for a meter. The City Council connects you to the mains, from where you extend a pipe to your house. The problem is many people are using poor quality plastic pipes, which are easily damaged, resulting in water contamination.

Like water, electricity is a lucrative business in the settlement. Some 20–25 households have electric meters, and extend their lines to others for a prede-termined price (based on the number of light bulbs or electric points used).

For example, one electric bulb costs the user KSh300 ($4) per month. "The profit from this business is 9–10 times the actual cost. For every KSh2,000 that you pay to the City Council, you make KSh20,000 ($285) from the community users," says the Chairman of the settlement. One problem resulting from these illegal or "shared" connections is the overloading hazard, which in the past has led to several fires and deaths.

Like many other informal settlements in Nairobi, the practice of "flying toilets" is common in Lunga Lunga. As reported by the community leader, "There are an estimated 50 toilets serving the 150,000 people of Lunga Lunga." Most of these toilets have pipes extending to the river, where the untreated waste is disposed. Further, there are no drains, and unpaved streets and alleyways in this tightly packed settlement leave little room for water to flow down to the river, thereby resulting in pools of stagnating water mixed with garbage and human waste.

All three of these settlements fit the standard definition of "slums." However, they are very different from each other in terms of how they were developed, and in terms of the specific problems they face. Gitare Marigu residents, for example, were moved to this area by the government, without any infrastructure, and forced to develop on their own. They did what they could do to improve their lot, but their "ownership" situation twelve years after a government resettlement program is still precarious. In the case of Lunga Lunga, people will need to be moved eventually—the site poses a hazard—but where and how is the question. In Kawangware, titles exist, but the rental business being run by the landlords has resulted in overcrowding and underservicing. In another, one of the more notorious areas, Mathare Valley, which had been developed by a land-development company, there were 8,026 people with only two pit latrines and three water points.[11]

Finally, there are many other areas in Nairobi where multistoried tenement housing is being built, in defiance of building regulations or basic living standards (lighting, ventilation, services). This business is a low-risk, high-return investment, for which there is increasing demand. As a result, rents are relatively affordable. Formal code-adherent buildings cannot compete with these costs and are therefore (we may speculate) not being built. The dynamic underlying the rental/tenement housing in Nairobi is worth briefly describing in more detail.

Recent work in the tenements of Nairobi shows how the system has worked in that environment. Four factors seem to have contributed to the situation. The first is that, since the 1960s, there has been a custom in Nairobi for people to get together and buy land for settlement. Soon these companies were effectively hijacked by the better-off, and instead of the self-built housing of the earlier days, row houses, closely resembling barracks and consisting of back-to-back single rooms without sanitation, were built for rent. The second factor is that landlords hide behind a complex screen of go-betweens, so as to hide their identity. It is well known that a large percentage of the upper echelons of the civil service are involved in these developments. A survey undertaken in 1984 found that when such units were built from temporary materials, the capital cost could be recovered in three years; if permanent materials were used, a period of seven years was required. At current rent levels a recent study shows a 100 percent return on capital within three to seven years within some tenements. This anonymity allows the landlords to use thugs to evict tenants who do not pay, and thereby limit the number of nonperforming units.

These landlords—who could also be called "developers" of this informal tenement housing—stop short of nothing to maximize profit. Building regulations are ignored, and restrictions on plot coverage, floor area ratio, and maximum height of the building are flouted. The resulting densities are truly amazing: in one area the density is 5,242 persons per hectare, which is over four times the density of the worst tenements in New York in the 1890s.[12] These tenements are characterized by insufficient natural light, nine to fourteen families on one floor, sharing a single toilet, and poor maintenance. The resulting environment, and the fact that the residents are all tenants with no personal stake in either the future of or management of the building, means that these buildings are depressingly unloved, and have every appearance of what we could call inner-city slums (see photos in Box 2.13).

Box 2.13   Tenement housing in Nairobi.

Source: Fieldwork by Mathema in 2005.

Having discussed informality in several African cities, next we present some non-African examples: low-income settlements in Afghanistan, Mongolia, the Philippines, and the United Arab Emirates. All have their very specific origins, character, and problems.

Kabul, Afghanistan

The informal settlements in Kabul present an interesting variation to some of the African cities. The informality results from the lack of basic infrastructure; most homes, despite decades of war, have secure (de facto) tenure. Densities range from 200 to 250 persons per hectare, comparable to that in formal areas. Site visits by the coauthor (Mathema) in late 2003 showed evidence of direct and serious damage in some settlements due to the war: traditionally constructed mud houses with plastic sheets being used to replace the collapsed roof and broken windows, for example—but at the time, none was "illegally" occupied, in the fundamental sense of the term.

Large families are a common feature, resulting partly from the culture of extended families (three to four generations living together) and partly due to the high birth rate (families with up to ten children spanning the ages of 1 to 20 years). Loss of male members, young early-teen girls not attending school, extreme poverty, and lack of basic infrastructure (water, sanitation) are some of the main characteristics of these settlements. Equally prominent is the hospitality and resilience of these families living in extremely harsh conditions. See Box 2.14 for snapshots of housing conditions in the informal settlements of Kabul.

Box 2.14   Informal settlements in Kabul.

Source: Field research by Mathema in 2003.

Mongolia

Mongolia presents its own unique building fabric. Large parts of the country's cities and towns are "informal" in the sense that they lack adequate infrastructure. But they all have de facto ownership. Plot layouts are fairly regular and well laid out. Plots are large in size, and densities are low. *Gers* or traditional Mongolian tents can be seen on most of these plots, often more than one on a plot, depending on the family size (see photos in Box 2.15): these are used as primary homes by the poorer households and as winter homes for those who can afford to build an additional structure for a summer house.

These *gers* are centrally heated using a coal furnace, which in general is a reasonably efficient and affordable solution, even for the poor, to keep out the cold. It is poor access to water and sanitation that is the biggest problem in these settlements. Water points are scattered around the settlements, where children come to fetch water in the freezing mornings with temperatures reaching minus thirty degrees centigrade (see bottom-right photo in Box 2.15). Each plot usually has its own pit latrine; and once full, it is sealed up and a new one is dug next to it. Most pits, however, do not meet basic hygiene standards.

Metro Manila, Philippines

In contrast to many of the examples discussed so far, the vast majority of Metro Manila's informal areas are squatter settlements. According to a Philippines National Housing Authority report, in the early 1980s, one

Box 2.15   Mongolia's *ger* areas.

Source: Field research by Mathema in 2006.

out of four Metro Manila residents was a squatter. In 2003, over one-third of the Metro Manila's population was living in squatter settlements on both privately and publicly held land. And even though many of the residents have lived in these areas for several decades, the very nature of their occupation of the land made regularization complicated, contentious, and politicized.

These squatter areas, commonly known as "depressed" settlements, match the typical characteristics of slums in other parts of the world: incremental construction, variation of building types, a high level of street activity, and inadequate physical and social infrastructure. Like other major developing-country cities, Manila's "slums" present a mix in terms of the quality of house construction: from temporary structures built with salvaged materials (wood and tin sheets), to permanent buildings built with concrete blocks (see Box 2.16). According to a survey in Metro Manila's depressed settlements in 2002–03,[13] strong and semipermanent houses comprised 86 percent of the housing stock in depressed settlements in all areas other than the hazard zones (creek sides, railway lines, etc.). In contrast, in hazard zones, the construction was predominantly temporary in nature, about 50 percent of the housing stock. The average density in these neighborhoods was 75 dwelling units per hectare, which translates into just over 500 persons per hectare (assuming a household size of 6.9, based on the same study).

Box 2.16 Metro Manila's "depressed" settlements (squatter areas).

Temporary houses made with "salvage" materials, mostly on hazard zones

Examples of semi permanent houses in squatter settlements

Example of a semi permanent house in squatter settlements

Source: Fieldwork by Mathema in 2002–03.

Dubai, United Arab Emirates

We close this chapter with an unlikely case, but one which emphasizes the need for development practitioners to look under the surface. This is the case of low-income settlements in Dubai, which presents a strikingly different visual picture from the conventional "slum" but in fact has deep-seated flaws which rarely get noticed. These settlements are not slums per se. These are formal residential areas built by the private sector for immigrant labor. So, unlike all the other cases we have discussed so far, these actually fit into the top-left square of Figure 2.1 with which we started this chapter, that is, into the "formal" settlements typology—in terms of their legal status and adequacy of services.

However, if one scratches beneath the surface, one will find that most of this labor housing in Dubai constitutes the sort of slum rental units discussed earlier, in terms of the plot density, room occupancy, the number of people sharing toilets, and so on.[14] The difference, however, is that they are built with permanent materials, just like any other formal sector housing (see Box 2.17). To an outsider, therefore, this might look like nothing out of the ordinary. But in fact, there are fundamental problems with this sort of arrangement, which is prevalent not just in Dubai but in all of the seven emirates of the UAE.

Box 2.17   Dubai's "labor camps."

Source: Field research by Mathema in 2008.

The sort of housing that is offered is illustrated by this advertisement from the real estate section of a local newspaper:

> LABOR CAMP in Sonapur. Brand new 162 rooms independent labor camp. With 72 Showers, 72 W/C, 3 Mess and 3 Kitchen, available for rent.[15]

The rent for each room is in the range of AED3,000–4,000 (approximately $800–$1100) per month, which usually excludes utilities. Private companies typically lease blocks of these rooms in one camp for their staff; some cover the cost of this accommodation but most do not. Rent for up to a year is paid in advance, when signing the lease. Given the high cost of accommodation, the only way for the target labor category to afford the housing is by sharing rooms: it is common for between five and fifteen persons to share a room. In the above advertisement, for example, this means 11 to 33 persons sharing one toilet and shower, and between 270 and 800 persons sharing one kitchen and mess (dining area).

Further important to note is the fact that this immigrant labor constitutes not an insignificant percentage of the population of the country but rather the vast majority. Here, a million-strong foreign workforce coexists with about a quarter of a million UAE citizens.[16]

Another issue is that, as evident from the terminology used—*bachelors' quarters* or *labor camps*—these areas house only men[17] with relatively low-paying jobs. These "bachelors"—because they come without family—are unwelcome in the "regular" localities occupied by "families," and accordingly, government policy requires these camps to be located at a sufficient distance from the main city. On the flip side, because this type of housing is the only choice for this labor class, it is beyond the realm of possibility for them to bring their families from the countries of origin. The social implications of this combination of isolation from the family, lack of any semblance of a societal structure, and alienation from the general community can be very serious. We shall not delve into those aspects here but rather leave the reader with a few pictures of this these "camps." Suffice it to say that the photos shown here are among the better-off camps; some of the worse ones were featured in a recent BBC documentary, "Slumdogs and Millionaires," with a vivid description of the appalling housing conditions—with raw sewage flooding the area, ill-maintained toilets, and most importantly, the face of economic desperation and social isolation—of construction workers in the UAE.[18]

CONCLUSION

There are many ways in which "the other half" can be identified, but the most potent is in the physical form of their housing, and here we draw a very important distinction between the slums of nineteenth- and

50   *Development Poverty and Politics*

twentieth-century Europe and the United States, and the informal settlements of the cities of the developing world.

As the quotation from Riis's book in Chapter 1 shows, most nineteenth-century slums were built within the formal system, typically
opportunistic subdivisions of older houses, converting them from comparatively acceptable tenements by building wooden walls to create ten
dwellings where one had been before. However, as the financial benefits
of slum building became ever more evident, in some cases they were built
then from scratch with no regard for the need for windows and ventilation as the plan in Box 2.18 shows.

There are common features between some aspects of these slums and
those of the informal settlements we encounter today. For example, for
some reason Nairobi is the home of some of the worst housing conditions
in Africa, and many of the characteristics of nineteenth-century slum-
lords have resurfaced in modern Africa, chief of which is the creative use
of limited space and resources to maximize income at all costs. This was
illustrated in the examples we used in this chapter.

Still, from the diversity of conditions evident, there is one simple point
to be made: one cannot generalize about the living conditions in low-
income settlements. The snapshots of housing and infrastructure conditions in lower income settlements across eleven countries—Swaziland,
Eritrea, Ghana, Ethiopia, Tanzania, Kenya, Nigeria, Afghanistan, Mongolia, Philippines, UAE—highlight the differences that exist not just
across countries and continents but also within each country and city.
In other words, the size of the city, the level of urbanization, the societal
culture, and priorities are all responsible for the prevailing conditions in
these informal settlements and need to be factored in when devising the

Box 2.18   Tenement of 1863.

(D: Dark; L: Light; H: Halls)

Source: Riis, Jacob A., *How the Other Half Lives*. New York, 1971.

solutions to the underlying problems. They also highlight that outward appearances can in fact be deceiving, that a settlement appearing neat and clean, as seen in the case of Dubai's labor camps, can in fact harbor more social and economic problems than the crumbling structures in the slums of Eritrea.

To stretch this warning against broad generalization on slums a little further, the next chapter provides some insight into the lives and lifestyles of the people who live within these settlements, and illustrates the hopes, aspirations, hard work, and most importantly the self-pride of these so-called slum-dwellers.

3   What Lies Beneath
A View from the Inside

Though informal settlements usually look like cities of despair and squalor, appearances can be misleading. This chapter tells the stories of residents in such settlements to illustrate not only the normalcy of their lives, but also their responses to the circumstances in which they live.

The last chapter discussed how these settlements and communities are very far from homogeneous, and, even where external impressions might be similar, deprived physical conditions often hide highly developed social ones. The many problems which residents face in informal settlements and slums are complemented by the achievements in the face of these problems. It is useful at this stage to look behind appearances and to try to understand the situation from the perspective of the residents themselves.

Here, we present stories of people living in these settlements. This could be seen as an attempt to romanticize the situation, but in our opinion no discussion on development, or poverty, or slums, or housing, would be complete without an understanding of not just what might be done but who we aim to serve. Ironically, the "who" often get relegated to statistics and numbers, or hardly goes beyond book covers, inserts, posters, and marketing material. There is much more. Who are these people? What makes them choose—or be bereft of choice—to live in such conditions? How do they cope with hardships and stigma? What is poverty in its real form, for instance, that goes beyond the generic description of earning less than a dollar a day (or more recently $1.25/day)? Who are we serving when we design development programs? What do they most need, aspire to, desire, or prioritize? What is the face of desperation and vulnerability, as opposed to hope and success? And so on.

We present stories of a broad mix of people. There are some who see their life in these settlements as a transition to something better, others for whom this is an improvement over their past situation, and yet others who choose to live in their present locations for purely social or economic reasons. There are some who are facing an extremely difficult situation, be it economic, social, or in terms of the physical environment

and infrastructure; others who have worked hard and endured, transforming their difficulties into opportunities or successes; those who
are in the process of this transformation; and finally, those who have
failed.

These examples do not pretend to try to answer the foregoing difficult
questions we ask. Rather, they seek to sensitize the reader to the real
issues facing these communities and provide the perspective and insight
from the inside.

Each of the persons interviewed (and many more, who have not been
presented here in the interest of space) spared two to four hours to share
his or her story, and our hope is that the reader will do justice to their gen-
erosity. But first, a few things to note: One, to preserve the privacy of these
individuals, their names have been changed and location-specific information (currency, city, settlement name) has been omitted. Two, all of the
illustrations are from Africa, and again, that is because that is where most
of the field interviews were conducted.[1] Three, some interviews date as far
back as 2004, but the story reads as current. That is partly to make the
presentation more consistent, but also because the intention is to illustrate
the situation more broadly rather than analyzing the specific situation of a
specific individual at a specific place or time.

The people stories illustrate the multifaceted role of informal settlements as places which: (i) provide a base for the young and old alike to
explore income-generating opportunities; (ii) provide the social safety
net necessary for people to cope with poverty; (iii) serve as venues for
entrepreneurship and creativity; (iv) present an alternative housing solution in the context of dysfunctional formal-sector housing markets; and
finally, (v) serve as a refuge of last resort for those who have nowhere else
to turn to. As these stories suggest, vulnerability and poverty are common features in these settlements, but by no means are all residents poor
and vulnerable. If there is one single common thread that runs through
all the stories, it is the factor of exclusion that the residents face—exclusion that bars them from assimilation into the formal sector, and exclusion from participating as legitimate citizens of a democratic society. It
is due to this very fundamental reason that the emphasis of this book is
on a development *process* that is inclusionary and people-centered, and
one in which these so-called beneficiaries and hitherto recipients of aid
are put in the driver's seat and allowed to lead the process.

Here, then, are their stories.

INFORMALITY AND VULNERABILITY

Mary

Mary is a 40-year-old single mother living with her three children, two
girls aged thirteen and nine years, and a boy of ten. She earns her living by

Box 3.1 Informality and vulnerability.

Mary is a 40-year old single mother, with 3 children. She bakes and sells local bread, and has a household income of less than $5 per month to spend on non-food items. They live in a 2-room rental structure built with mud.

Joyce (right) is a widow with 7 young children, and the caretaker of her 85-year old mother (left) as well as 7 more children from her deceased sister. She runs a small vending business for a living. The 16-member household occupies 2 small rooms of an ancestral house that belongs to the family.

 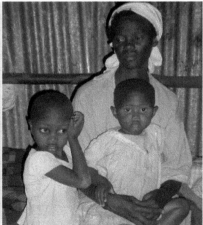

Vida is a food vendor, the main bread-winner of an 8-member household comprising her husband and 3 children, all of whom are currently unemployed, and 3 grandchildren. They live in an incrementally built 2-room structure located in a flood-prone area on the beach.

Fatima, married at 14, was widowed at 30 years of age, and has 4 children. Her husband died of AIDS recently. She sells charcoal for a living, from a small stall outside her house, and the income is about $60 per month.

baking and selling the local bread, income from which is a little over $1 a week, or $5 a month. Food comes from her business, so this $5 is used for other expenses. Mary moved here over thirty years ago as a child, and has lived here in a government-owned rental unit since. The rent is subsidized, about $0.80 per month, but she has not paid for over five years, and hence has an outstanding balance of about $60.

Mary's house is roughly 4 m × 5 m, divided into two rooms: the front room houses a clay oven and is used primarily for cooking. The rear, windowless, space is the sleeping area. She says she cannot borrow money from the government's microcredit facility because that requires a guarantor with a salary of over $100 a month, and there is no one she knows with that sort of income who will support her.

She has her own piped water supply, recently installed by an NGO, and pays about $1.5 per month for water. This excludes the $6 that she yet has to pay for the pipe installation. Her family shares a pit latrine with five other households in the vicinity. That is filled up now, and the landlord has declined to get it emptied as the users have no money to contribute.

For medical treatment, she uses the public health facility, which is free, but she suffers from a chest problem, very likely tuberculosis, for which she has to make frequent visits to a private hospital. Two of her children go to a government-subsidized school, which is a thirty-minute walk from the house. While tuition is free, uniforms and books cost about $1.5 per month.

Joyce

Joyce is a widow who lives with her 85-year-old mother and fourteen children: seven of her own and seven from her deceased sister. The husband of the deceased sister is a fisherman and lives nearby; he provides occasional child support for his seven children. With so many members in the household, and only two rooms, several of the children go to the neighbor's next door to sleep at night. Joyce was born and raised in this neighborhood and community. Apart from her married years when her husband was alive, she has always lived in this house.

Joyce runs a small business with her mother: they buy pigs in the rural areas, bring them to the city for slaughter, and cook the meat and sell it locally. They do not have space or the money for a refrigerator, so they typically store the extra meat in a neighbor's fridge for a fee. Apart from income from this business, they earn some money by selling water from their private standpipe. The total household income is about $80 per month.

There are several semiconnected housing units on the plot. The main building is single story, made of permanent materials. The four rooms in that structure are occupied by four other households. Joyce's family lives in an adjacent block made of plastered mud blocks. A small outdoor space serves as the washing/cooking area, and also has the water standpipe. Joyce would like to have a bigger, better house, but her primary concern is the lack of a toilet. She says, "Renovation/improvement/additions to the building require too many approvals. The process is not only time-consuming and difficult, it is also very costly. In addition, the cost of construction is high, and we don't have that kind of money to invest."

Joyce says there is no official document to prove ownership of the individual structures, but the head of the family does have a title deed to the plot. He does not live here, but pays property tax, towards which Joyce contributes her fair share. Her assumption is that the property is "registered" *since* they pay taxes.

They have an electricity connection, but the service has been disconnected for the past six months due to nonpayment. There is a water standpipe in the plot, but the supply is not regular. There is no toilet in the compound. The families use the public toilet provided by the local municipality, which is connected to the main sewer lines. It costs $0.10 per use. Assuming even one visit per day by each member, this would add up to some $35 per month. Joyce says the children often use the school toilet, and the adults use the facility once in 2 days or so, so they don't spend that much. Still, she says, "It is extremely expensive. When the colonial masters were here, it was free. Now this government wants to charge a fee." The public shower also costs money—about $0.07 to $0.10 per use. "We don't use the showers; we just shower in the yard here."

Sixty-five percent of the household income is spent on food. Water takes up another 7 percent and public toilets 9 percent. They try to save about 20 percent of the income on a regular basis. Joyce has a *susu* (informal savings club) account which she contributes to on a daily basis. At the end of each month, she withdraws the money and invests it in the pork business.

Joyce's mother says, "If we weren't poor, we would have been able to pay our basic bills without help from others. Still, I can say we are better off than many others here: I am poor but I have respect in the community."

Vida

Vida lives in a small house on a beachfront in an area zoned as a hazard zone, about one hundred meters from the water. She is a food vendor, who cooks and sells traditional food. She is the primary breadwinner of a family comprising her husband, three children, and three grandchildren. Her four older daughters are married and live elsewhere in the city. Vida's income is about $135 per month. Her husband is a fisherman, but has been sick and bedridden for several months and, therefore, unemployed. Her children are also unemployed except for her 21-year-old son who works occasionally at a car wash, earning about $15–$20 per month on average. This makes the total monthly household income average about $150.

Before moving here two years ago, Vida lived in her father's house in a nearby settlement. But due to a family dispute, her household along with her mother moved to this area. Her mother, who died a few weeks ago (Vida was in mourning at the time of the interview), had "bought" this plot from the previous owner for $110 five years earlier. Vida has no receipt or proof of purchase, but says, "It was sold to us by the owner on good will . . . he was a family friend. However, the plot is so small that even if I try, I don't think we will be issued a title. But it's something we can make do with until we are asked to move." When asked if she understands that zoning regulations might not permit living so close to the water, she says, "Most people close to the water are fishermen, with boats, etc. They don't really care if it is legal or not. It's convenient. Proof of ownership lies in the fact that the people still live there . . . and no one has asked them to move out."

"We are much worse off than most other households here. My husband is sick, none of my children are employed, and I lost my mother a few weeks ago, due to which I had to incur high expenses for the funeral. Because of that, the business is not doing very well either. These are hard times," says Vida.

The house has two rooms, and another one is under construction, to be used as a cooking space for her business. It is being constructed incrementally by a friend who is a mason. The walls are made of concrete block, the roof out of asbestos sheets. There is no toilet or kitchen. There is no electricity or water on the plot. Vida buys water from her neighbor, spending a total of about $10 per month on water. With no toilet facility, the adults in the family use the beach early in the morning; at other times, they use public toilets, paying $0.05 per use. The children defecate on the beach.

Vida spends about three-fourths of her income on food. About 10 percent goes to purchasing water (7 percent), and using public toilets (3 percent). She has no savings, or any savings account. "I have heard that people go and wait all day at the bank, and get nothing in return," she said. Her son added, "With so little money, it is even difficult to buy in bulk and increase the profit from the business, let alone save."

Fatima

Fatima is a single mother with four children: a ten-year old boy, and three girls, fourteen, four, and two years old. Fatima was widowed last year, and now manages the household expenses with a small charcoal-vending business which gets her a profit of $15 per week (selling one bag of charcoal per week). Time permitting, she sometimes sells vegetables, with assistance from her eldest daughter on the weekends, to generate additional income.

Fatima came from her village to the city at the age of thirteen to visit her grandmother living in this settlement. She joined the local primary school and studied there up to grade 4. A year later, a catholic church was built in the place where Fatima's grandparents lived, and so the family was asked to vacate the land. Fatima dropped out of school and took up part-time work as house help in a neighboring high-end area. There she met her potential husband and got married at the age of fourteen.

Fatima and her husband took up rental accommodation in a nearby settlement; she was pregnant at the time with their first child. Her husband got a job as a cook in a hotel nearby, earning a handsome salary of $45 per month. The room rent was $12 without any water or electricity and, at the time, was affordable. They lived there for four years. When she was pregnant with her second baby, they moved to the current house. They found it at the same price as the old unit, but much better in terms of location (on the main road) and more spacious. Fatima's husband got sick with tuberculosis last year, moved to his home village, and died there. According to a local health worker, he had AIDS, and Fatima is likely to have contracted it too, something which Fatima is unlikely to admit, and understandably so, given the stigma associated with it.

The housing structure is a single-story series of rooms of which Fatima rents one. There are three toilets on the plot, shared by some twenty households, totaling some sixty to seventy persons, all renters. Each renter is given a key to one allocated toilet, and maintenance is a common responsibility. The landlord gets the pits emptied when they fill up. Fatima purchases water from a vendor, which costs her roughly $6 per month. There is no organized system for garbage collection; it is disposed of on the streets/corners. There are no drains or drainage system within or outside the plot.

Her 14-year-old daughter and her son attend a public school, which is a twenty-minute walk from the house. The tuition is free, but she needs to pay for the uniforms and an annual examination fee of about $11 per child.

Fatima's says her difficulties have increased since her husband's illness and subsequent death. "The rent is too high, and I haven't paid for five months. The landlord came knocking today, but I explained my situation. He has warned that I will have to move out if I can't pay by next month. The only advantage of this house is that it is near the roadside, where I can do my charcoal business while watching the kids. But still, it is very expensive. With such little income, and four young children to feed," she says, "hunger is the biggest problem. There are many days when they go without food."

Fatima is not a member of any women's organization or community group, and not aware of any assistance programs that can help her. When she doesn't have money to pay for groceries or other things (water, etc.), she takes them on credit and pays when she can. She wants capital to invest in a larger business, she says. "If I could have a big business, maybe I could earn more money and my life would change." Asked whom she would categorize as "better off," she said: "Those who have TVs, stone houses, and surplus money, i.e. those who don't have to live hand to mouth, those who have enough food to eat."

Box 3.2 Informality and income-earning opportunities.

Ester is a divorced, 28-year old mother of two young girls. The older daughter lives with Ester in her 1-room wooden kiosk in the city, the other one with the grandmother in the home-village. Ester is a new resident in this settlement, and barely making ends meet by working as a seamstress.

Michael is a 25-year old who works as a food vendor, and shares a room for rent with two friends from his home-town.

Jamal moved out of his parents' house some 10 years ago, rented a room, and started his own small business. Today, he makes about $80 per month trading in cell phones and clothes.

Margaret is a 29-year old who lives in an informal settlement near a railway track, sharing a room with 3 other female roommates. Joyce is a trader; she sells biscuits.

INFORMALITY AND INCOME-EARNING OPPORTUNITIES

Ester

Ester is a 28-year-old single mother of two young girls. Ester works as a seamstress for a living, taking small orders in her house, which also serves as her sewing workshop. Income from this business is irregular, she says. "I get $4–$5 per dress, and I do between three and five a week, so on average I make $80 per month. But oftentimes I get few or no orders, and I barely make $40–$50." Her brother, who lives nearby, provides monetary assistance occasionally, approximating $10 per month.

Ester is the second oldest among six siblings; her father had a cocoa farm in the village. She stopped schooling when she was very little—in grade 1—due to financial constraints. When she grew up, she worked as an apprentice for a tailor for three years.

Ester was divorced six years ago. Her younger daughter, six years old, lives with her and attends private school here; the older daughter, nine, lives with Ester's mother in the village. The ex-husband does not provide any child support or alimony. She came to the city two years ago, with her younger daughter. She was invited by her brother, a porter/bar owner based in this settlement. He helped her find a small piece of land and build a kiosk for shelter. She paid $30 for the land to the "owner," a neighbor of the same tribe. "Tailoring is the only thing I know how to do," she says, "so I decided to start up that business here. I have to raise my kids somehow. My brother used to pay for my children's schooling until a year ago, but now he has his own children to feed, so he cannot anymore."

Ester's house is a 3 m × 3 m temporary structure made of wood and tin sheets. Ester uses one corner of the room as her workshop. They often use her brother's kitchen, next door, and eat with them when they are short of money. Asked if she knows how long she can stay on this land or if she has any documentation for her land transaction, her response was, "There's no defined period really. Presumably, I can stay as long as I need to. As for proof of ownership, there is none. But since the seller is also living on someone else's land, and he is from my tribe, he can't kick me out. I have heard about the government's plans for clearing this area and resettling the residents. I am hoping that I will make enough money before then to be able to rent another place somewhere else. As for the money I paid for the land . . . well, I can't really make any claims or challenge the government because I know the land belongs to them."

Ester has an electricity connection. "It's very difficult to identify the meter," she says. "There are many of them, and they are all interconnected. Someone comes to collect the money. . . . I pay a $2 flat fee every month." She buys water from a neighbor, for which her monthly expense is about $10. She uses a private pay-to-use toilet which costs $0.05 per use, amounting to about $3 to $4 per month. Public showers cost $0.07 for adults and

$0.03 for children, which is an additional cost of about $4 to $5 per month. There is no organized system for garbage collection or disposal. Trash is disposed in the lagoon adjacent to the settlement. "There's no garbage container," she says, "so I have no choice." Wastewater is disposed of in the unpaved street in front of the house.

Ester spends half her income on food, and about 20 percent on services (2 percent on electricity, 10 percent on water, and 8 percent on public toilets). Another 8 percent goes to her older child's education in the village. What she has left she saves in a *susu* account, roughly $1 per day. "I don't own any land or property," she says, "so it's been difficult so far. And even though I am in a difficult situation, I think I am better off than many others. At least I have a business. Many have nothing." She defines better off as: "being able to save for an emergency, as I am able to do in the *susu* account."

Michael

Michael is a 25-year-old who makes a living by making and selling traditional peanut candy and freshly brewed coffee. He makes his sales going house to house carrying the goods in a basket on his head. From this work, he earns $1 to $2 a day, averaging to about $40 per month. Michael is originally from a farming family in another town. He is an orphan, and lived with his grandparents in his hometown until 2006, when he migrated to the city in search of work. He has been living here since, sharing a one-room rental accommodation with two other friends from his hometown. Michael is the oldest amongst his siblings; he has never attended school because the family could not afford it. Now he works and sends money home to support their schooling. He sends about $1 per month, and visits them two to three times a year: the bus ride there takes five to six hours.

The main building in the plot has three rooms, built of concrete blocks and roofed with corrugated iron sheets. One room is occupied by Michael and his two roommates and is very modest: about 3 m × 2.5 m in size. The rent is $10 per month, which is split equally among the three roommates. There is an open space and a small shed outside, which is used by the residents for cooking and other outdoor chores. Michael and his roommates prepare their traditional sweets and coffee there before going out to sell it. With regard to the title or legal ownership of the property, Michael says the owner probably has paperwork for the property but is not sure. The chairman, on the other hand, said that this property is in a flood zone, so it is unlikely that the owner will get a title.

The house has an electricity connection for which Michael and his roommates pay a flat amount of $1 per month, which is included in the rent. Charcoal is used for cooking, which costs Michael about $2 per month. Water is purchased from a neighbor. There is a flat charge of $1 per month for unlimited usage, which is split among the three friends. There is a pit latrine in the plot along with a bath area. Water from the bath is channeled into an open drain, which culminates at the access path. With no outlet, the wastewater stagnates there along with other garbage, making it a breeding ground for mosquitoes. Michael expressed his concern for the drainage problems around his house. "It's a health hazard, and we all know it, but there's very little we can do. The water has to go somewhere . . . and because there is no outlet, it sits here. It gets particularly bad in the rainy season." As a result, malaria is very common, he says. He contracted it a month prior to the interview, and the treatment cost $25, which is substantial. Garbage is picked up every two to three days by a private collector. The fee is $0.3–$0.5 for each collection (depending on the size of the bag), which again is split among the three roommates.

Michael has no savings account. "I am interested," he says, "but I have no savings. What I save, I send home to my family." Food takes up to 73 percent of Michael's income. Another 9 percent goes towards house rent, 9 percent to medical bills, and 6 percent on services (1 percent each to water, electricity and garbage disposal, and 4 percent to charcoal).

He is interested in a loan, he says, but not for a house. "I can't afford a house right now. I first need to make some money. I would borrow to expand the business." On savings and credit cooperatives, and why he does not get together with his roommates and apply for a loan, he says: "For an individual, they require collateral (TV, radio, other assets) or savings to qualify for a loan, and I don't have either. A group loan is difficult because people are scared; they don't trust each other. What if one member of the group runs off? The others are stuck paying off his dues."

Jamal

Jamal is a 28-year-old single male who works as a trader. He is a member of the local youth association and lives in a single-room rental unit in a courtyard house. Jamal's parents live in a family house in the neighborhood, which he left ten years ago to take up this rental accommodation. Jamal trades in clothes and phones which he brings from across the national border and sells locally. From this, he earns about $850 annually from his trading business, that is, $70 per month on average.

The plot has twelve rooms lined around a courtyard; eleven are rented, and one is occupied by a caretaker. The landlord lives elsewhere. The walls are made of concrete blocks, and the roof out of tin sheets. There is no kitchen or toilet. Jamal does not cook, but the other renters cook in the courtyard. Jamal occupies one room for which he pays $9 per month. He pays twelve months' rent in advance and renews the lease every year. Three years ago, the tenants contributed $3 each, and paved the courtyard. "We asked the landlord to do it several times," says Jamal, "but he kept postponing it, so we decided to do it ourselves. I would like to have my own house some day, but right now I have no money. Land is very expensive, and houses are mostly targeted to the rich. I do have my eye on a plot of land, but for which I am saving up money."

There are three electricity meters on the plot, and the bill is split according to the number of electrical points. Jamal's average monthly bill is about $3. Water is bought from a neighbor. Jamal uses the public pay toilet, which costs $0.04–$0.05 per use. Jamal has built a small bath area within his room, which drains into the street outside. He says, "Currently, we have a plan in the pipeline to build more public toilets and showers in the settlement, but are awaiting permission from the municipality."

About 45 percent of Jamal's monthly income is spent on food, another 13 percent on services (water, sanitation, electricity), and 15 percent on rent. Jamal has had a savings account in a bank for five years now; his current balance is about $1,000. He saves $15–$20 per month on average. He does not trust the *susu* collectors, he says, "because they run off with your money, and one can't even trace them."

Margaret

Margaret is 29 years old and shares a single rented room with three other young women near a railway track. They are all petty traders, who walk around with baskets on their heads, selling goods. She sells biscuits. The four girls cook together for the most part: they pool money for food, which they share. All other expenses are separate. Both of Margaret's parents remarried, and have one child each. She is not in touch with them, but she is in regular contact with her grandmother, who lives in her hometown and sends her money when she can.

Margaret first came to the city some four years ago with an aunt to help with her trading business. She lived with the business owner in a "company house." After a year and a half, the relative returned to the village, and Margaret took up another job at a small restaurant. She lived rent-free at her workplace, which had a bath area but no toilet. She worked there for a year, and then moved to the current settlement to work as an assistant in a small snack bar. There again, she lived with the business owner, in accommodation similar to the previous one. After a year, she met her current roommates and decided to move in together with them. One of her friends was a biscuit seller, so Margaret decided to quit her job and go independent; she started the same line of business. She now earns some $135 per month on average.

Margaret's house is a one-room structure, 3 m × 4 m, with brick walls and a tin roof. It has a little covered shed in front, which is where the girls cook. Adjacent to the structure is a large open space, where garbage is often dumped and burned. "The smoke that comes from there is really unpleasant," she says. "Many times, the butchers bring their dead animals and burn them using rubber tires to take off the hair. The rubber produces thick black smoke, and worse, it smells terrible!" (Incidentally, we witnessed this firsthand during the interview.) The owner of this house does not live in the area. They were living rent-free until five months ago, when he decided to start charging them $6 per month. They are now required to pay four months' rent in advance, every four months, that is, $24 every four months. "Flooding is a problem here. You can see the stagnant pool of water right in front of the house. And the smoke from the burning of animals. . . ." When asked why she still chooses to live here, she said: "We weren't paying rent here initially, so that was a saving factor. But now we are, and I think I would like to find something else."

The house is not fitted with electricity. "We couldn't afford it even if we had it, so it's okay," says Margaret. She spends about $2 per month for water for cooking, which is purchased from the neighboring bathhouse. Margaret and her roommates use the pay-to-use shower and toilet facilities. The garbage is usually dumped in a municipal container. "The man there always demands money per bag of garbage disposed," she says, "but I always trick him and manage to get away without paying!"

Margaret spends a third of her income on food. The bath and shower are more expensive here than some of the other comparable settlements in the city, and she spends 11 percent of her income on that. She is able to save about 30 percent of her income. Margaret has no bank account, her main concern expressed thus: "What if I move from here? Will the bank return my money?" Apart from a *susu* account to which she contributes $1 everyday, she was not aware of any systematic savings schemes that she could be eligible to participate in.

SOCIAL SAFETY NETS IN INFORMAL SETTLEMENTS

Irene

Irene is a 34-year-old single mother of two young girls aged thirteen and seven. She was divorced from her husband three years ago, who now provides child support of $1–$2 per day, about $50 per month. Irene sells vegetables in a nearby market, where she owns a stall outside her brother's house. Her income from this business is about $40 per month. She is also the secretary of the local government subward. There is no regular salary for this job, but she gets an allowance equivalent to half of the total fees collected from the residents, which amounts to some $25 per month on average. Her total monthly income, therefore, is about $115, plus an occasional $5–$10 from family members in the form of assistance, if required.

Irene is originally from another town. She got married in 1992, and moved to the city with her husband. They lived in a rental room for $4 per month until three years ago when they got divorced. It was then that they found this piece of land through an agent, which they bought for $370 with part of Irene's inheritance money (some $850). The remaining money was

Box 3.3   Social safety nets in informal settlements.

Irene, a single mother, is a vegetable vendor, and also the sub-ward Secretary. She is building her house incrementally and as she does that, her neighbors allow her free access to their toilet facility as well as water point.

Carmen is a food vendor. She lives with her mother and two children as part of an extended family of 16 in a single room of their ancestral family house. It's crowded, she says, but at least they have a roof over their head.

Agnes, 68, is head of a 15-member household. They live in a 5-room rental accommodation, and pay nominal rent to the landlord. The landlord is a friend, and thus lenient, otherwise they would not be able to afford this place.

Rose, a 28-year old, is divorced, and has one child. She runs a food stall along the railway tracks, and lives in a 1-room wooden enclosure. She supports her three younger siblings and also is the caretaker of her sister's 3 children, all of whom live with her in her 1-room kiosk.

put into the construction. "I moved into the mud house that was on the plot originally, and began construction of the house. I completed one room of the three, and the money finished. So the incomplete rooms have been sitting like that since." (Three years after the interview when I met her again, the condition of the house was the same.)

Irene's house is made of compressed mud bricks. Irene's family occupies the one room that is complete; the other two do not have a roof or windows yet. The house does not have electricity; Irene uses charcoal and kerosene for cooking and lighting, which cost about $17 per month.

Irene gets water from her neighbor. "It's a deep well from where water is pumped up and drawn from a standpipe. They typically sell the water," she says, "but I don't pay because we are friends." Asked whether she thought the water was fit for drinking or whether they treat it before drinking, she said, "We don't boil it because it doesn't taste good like that." There is a modified pit latrine in the neighbor's plot which Irene's household uses. Here again, she does not pay for daily usage, but contributes $6 towards the annual clean-up (which costs a total of $15). There is also a bath adjoining the pit; wastewater from the bath goes into the same pit as the toilet, while that from washing and so on is simply tossed into the street. Garbage is picked up by a private collector who takes it to the garbage dump a little distance away. He charges $0.60 per month.

Both her daughters attend public primary school, which is within walking distance from the house. They take private classes after school, which costs a total of about $10 per month. In addition, books and uniforms and so on cost $2–$3 on average, making the total monthly cost of education roughly $15. The medical facility is within walking distance of the house. Irene's annual medical expenses are in the range of $10.

Irene is a member of a local savings/credit group to which she contributes between $1 and $2 per month. Her current savings balance is about $45. "It's like an NGO," she says. "We save a little, and then are eligible to borrow. I borrowed $42 from them in September 2006 for three months. We had to repay $4–$5 per week, for three months. Before qualifying us for the loan, they would come to our houses, and value the assets, like the furniture, TV, and so forth" This was Irene's first loan, and having paid it all off, she qualified for a larger loan of $130, which is current. Irene has no other savings account, but plans to open two accounts for her two daughters at a "proper" bank, where she can start saving for their future.

Carmen

Carmen's household comprises her mother and two children, but she shares her accommodation—one room in the ancestral family house—with her four sisters and their children, totaling 16 persons. Carmen is married, but her husband lives in his own ancestral home for lack of space. Only three members of this 16-member household are males, and all of the seven other households on this plot—Carmen's cousins and their families—are women-headed. Combining all the households living on this plot, there is a total of nine males (among whom only two are adults), and 31 females. Carmen was born and raised in this settlement and has lived in this house all her life.

Carmen runs a small home-based food business. She cooks traditional porridge in a temporary shed outside the house and takes it to the street side where she sells it from a stall. She is assisted in this work by her sisters and shares the profit from the food vending business with them. The total household monthly income is about $120.

The house is a two-story structure, located adjacent to the local chief's palace near the main access road into the settlement. The lower level is built with permanent materials, and the upper story is made of wood and tin sheets. Carmen's sixteen-member household occupies one room on the ground level. The room is mainly used for storing their belongings; they sleep outdoors. The head of the family, in whose name the plot is, lives in another settlement in the city. Carmen is unsure of the type or terms of the title or registration. "I would like to move. There's no point trying to make the improvements here . . . too many complications. Once I make enough money, I will build a house elsewhere. I would be interested in a loan, but I don't have any documentation or collateral."

There is a common electricity meter in the plot; the bill is shared by all the resident households, based on the number of electrical sockets they have. There is no water connection on the plot. They buy water from a neighbor. "We know we are paying much more than we should be, but getting an individual water connection is very expensive," says Carmen. There is no toilet in the compound. The families use the public toilet provided by the municipality, which costs close to $0.10 per use. Carmen estimates that her household (16 members) spends about $20 per month on toilet usage. "I would like to build a toilet. Having your own toilet, however basic, is better than using a public toilet. But our grandfather did not build one when he constructed this house. Now there are too many owners of the plot which complicates any possibility to reach any agreement on renovation decisions."

The family spends 45 percent of the income on food. Two percent is spent on electricity, and 5 percent each towards water and sanitation. Carmen has been a member of a savings scheme for nearly a year now. She has already taken two loans to expand her business: the first was about $110, which she has paid off, and the second is $250 for five months, which she is currently repaying, She has managed to save some $50 in her account, and continues to contribute to this fund so as to be able to pay for her youngest child to go to school. In addition, she contributes about $1 per day in a *susu* account.

Agnes

Agnes is an unemployed and sick widow with a large family. She rents five of eight rooms of a large house. They have been living here for over fifty-five years now. She moved to this house with her husband when they got married. Agnes had 14 children in all, six of whom are now deceased. Two boys died in a motor accident, two more died in a shoot-out, and two died due to some illness. She is now left with eight, five of whom live with her, along with many grandchildren. Agnes and the grandchildren (eight residing here) share two rooms along with her 36-year-old single daughter. Another room is occupied by two of her single adult sons. The other two rooms are occupied by the other two sons and their spouses. In all, there are 15 residents occupying these five rooms.

Two of Agnes's sons and one daughter-in-law together make some $300 per month and contribute about a third of this toward household expenses. Another son lives in the United States and remits $200 every six months or so, roughly $30 per month. The total monthly income, therefore, is about $140.

Part of the structure is built with unplastered concrete blocks, part with wood, and a tin roof. There is a courtyard where most of the household chores (cooking, washing) are done and the cows are kept. Agnes's household pays a rent of $20 per month for the five rooms. When they first occupied the place 55 years ago, they were paying less than $1 per month. Agnes claims they had a good relationship with the original owner, who is now deceased. His family lives in the central city. "They own a lot of properties, and are managing this one as well. They originally had three houses in this neighborhood. They sold off two, but have let this one be because of my poor health. They know we are poor and I am sick, and can't go anywhere else, so they let us stay."

Agnes also owns another plot of land (with a registered title) in an area some 10 km from this neighborhood. "But," she says, "we have no money to build a house there. We've built a one-room structure, where someone stays to watch the land."

There is no electricity connection in the house. Agnes's family buys water from their neighbor when the water supply is good. At other times, they hire a taxi and bring water from another area. There is no toilet within the compound. The family uses public toilets, paying the equivalent $0.04 per use.

Nearly two-thirds of the household income goes to food, about 16 percent on the house rent, another 14 percent on public toilets, and 7 percent on water. That leaves practically no savings. One of her sons has a personal *susu* account. "He had also tried opening an account with a local bank, but too much documentation was required, so he dropped the idea and went with the *susu* collectors," says Agnes. Asked if she or her family knew about the informal sector savings schemes, her son responded: "Yes, I learnt about one from a friend. It was a pyramid scheme, and turned out to be a scam after all, because of which people don't trust these schemes anymore."

Rose

Rose is a 28-year-old food-stall owner, a single mother with one child, who lives in an informal settlement along a railway line. She is separated from her husband. Her child lives with her mother-in-law, but she is the caretaker of six other dependents: three younger siblings 17, 12, and 10 years old), and three infant nephews (2-year-old triplets) from her sister, who lives in another city. The sister is unemployed, and has four more children (seven in all), and Rose is lending a helping hand. From her business, Rose earns $7–$10 on average per day, about $230 per month, of which she sends $20 every month to her parents in her home town.

Rose came to this city after she got married, ten years ago. She and her husband first lived in a rental unit in another settlement, and it was then that she started a small food stall in this area. After a few years of marriage, she got separated from her husband and moved to this railroad track. While her stall is right next to the track where most of the activity occurs, she lives in a wooden shack on the other side of the settlement near a canal. "This is certainly not the best place to live," she says. "When it rains, the area gets flooded, which hampers the business as well. But there's no other place to relocate to. If the government allocates me a spot somewhere, I would be very happy."

Rose's house is a single-room wooden kiosk, roughly 3 m x 3 m, which accommodates all seven residents. She has no title to the property and is unlikely to get any since this is zoned as a hazard zone (very close to the railway line). Rose says the municipality came and demolished her first structure which she earlier used as her food stall, but she rebuilt it. She now lives in that and uses a mobile cart for her food stall. "This time," she says, "I got verbal permission from the owner of the big building behind this house. Some time ago, another individual came to evict us [the entire community], claiming this was his property. The case is currently in court. Officials from the municipality come as well, every few months; they mark our structures with red crosses, implying that they are to be demolished, but in the end we [the community] pool our funds and give them a levy, some $4–$5 per year, and they let us be. More recently, we have been hearing more threats, but if they try to evict me, I will plead with them to let me stay. I have nowhere else to go, and six mouths to feed."

Rose has got an electricity connection from the neighbor's house, for which she pays a flat rate of $6 per month. She buys water from a neighbor, which costs her about $25 per month. The family uses a private pay-to-use toilet which costs $0.13 per use, the monthly expenditure for which she estimates at $15. Public showers cost $0.11 per visit, roughly $12 per month. The total expense on toilets and showers is, therefore, $25–$30 per month. There is a garbage skip provided by the local municipality

Box 3.4   Informality and opportunities for entrepreneurship.

Kofi runs a pay-to-use toilet and shower business in an informal settlement from which he earns about $1,000 per month.". . . . I knew I could start something new and innovative," says Kofi. "I observed the needs of the market women here . . . . there were no baths or toilets. So I decided to start that business."

Steven runs three home-based businesses: water sale, pay-to-use public showers, and an electronics retail shop. His income is about $340 per month.

Florence runs a little restaurant, has 3 employees, and makes about $200 per month. Most of this she sends back to her family in the village. "I am in the city for the business, to make the money," she says. "There is no room for the kids here; they are better off attending school back home. I visit them occasionally."

nearby, where she disposes of the garbage. The supervisor of the skip charges $0.05 for every bag of garbage. "It's probably illegal to charge this fee," she says, "but we pay anyway, just to avoid being insulted or treated badly."

Expenditure on food is minimal since Rose's family eats at the food stall. However, the expense on services is high: 2 percent of the income is spent on electricity, 11 percent on water, and 12 percent on sanitation and waste disposal. The bulk of her income, however, goes to a savings-and-loan scheme. Rose is a member of a group-savings scheme, where she has a loan of $220. Her group has 22 members, and she contributes $3 every week as a compulsory savings deposit. She got this loan within a few weeks of joining the scheme, which she is slowly investing into the business. "The representatives from the scheme give us advice on how to use the money," she says.

INFORMALITY AND OPPORTUNITIES FOR ENTREPRENEURSHIP

Kofi

Kofi is a 48-year-old business owner who lives with his wife and six children in an informal settlement. He operates a toilet/shower business which is very profitable, generating an income of about $30–$35 per day. His wife is a vendor, who sells drinking water sachets; from this, she makes on average $2–$3 per day. He also owns some 20 goats, which he keeps in his yard and sells as a side business. Although not an elected leader per se, he commands respect in the settlement because of his "entrepreneurship," he says. Four of his children attend school: one is in a technical vocational school training in construction, and the younger three attend private school, which is expensive, but he is managing with the income from the toilet/shower business, which is in the range of $1,000 per month.

Kofi's home town is twenty-four miles from this city. He came here after finishing school in 1985. He was a musician and performed in a band for three years. He then met his present wife, got married, and had two children. Soon after, he got a music contract in Liberia, where he spent a year. In 1990, when he returned, he says, "Things were very difficult. The music business wasn't fetching enough money, so I quit and took up driving. My brothers loaned me two cars to start my business. There was more money in driving, but I wasn't happy. So I quit yet again."

He then inquired with the "owners" of this area if he could get some land on the other side of the settlement, near the lagoon, which was empty at the time. He paid close to $900, and was allocated a piece of land. There was no title, he says. "It was a verbal understanding that we had a 99-year lease. I have a receipt for the payment." There he built himself a small one-room structure. "It was then that I knew I could start something new and innovative," says Kofi. "I observed the needs of the market women here at the time: there were no baths or toilets. So I decided to start this business." He started out with five showers and slowly expanded his business to include pay-to-use toilets. "Building toilets was not an easy task. I had to obtain permission from the municipality, since there is a health issue involved with toilets. They first refused permission, but then I brought them here to see people squatting [defecating] by the lagoon. It is then that they permitted me to build public toilets. It was tacit approval—no written documentation was provided of course."

Kofi's house is built of concrete blocks and has a tin roof. There are two structures, with four rooms in all. Two more rooms are under construction, which he plans to rent out. He started out with one room, and has slowly added more rooms over the years. Cooking is done outdoors. He owns another plot of land on the outskirts of the city, which he bought from the local chief. This is a 0.06 hectare plot for which he paid about $1,700 and got a 99-year-lease deed. He has a government-issued title, but he says, "There had been multiple sales for the same plot . . . When I started to construct his house, another 'buyer' came and demolished it. The property is now in dispute. And although I have the title, it seems the land is not registered. How can that be? In any event, I know I am the rightful owner to the property. I can take the other claimants to court, but who has the time to do all that? Besides, these things take years to resolve. So instead, I have decided I will go to the other party and pay them something to make peace and resolve the issue."

Kofi spends about a third of his income on food. His other major expense is education for his children, which takes up about 20 percent of the monthly income (the technical school costs about $400 annually, and the private school for the three kids about $1,000 for tuition plus $500 towards boarding and living). After all the expenses, he is able to save about 30 percent of his income. He is not interested in a loan because, "it is too complicated. Too much documentation is required," he says.

Steven

Steven lives in a beachside house in an informal settlement with his family: a wife and five children. Two of these children are his sister's, whom he is taking care of due to her financial difficulties. The household occupies two rooms in this house. Steven moved to this settlement in 1986, and bought this plot of land (12 m × 24 m) from the so-called caretaker of the land. He built one room to begin with, and then after 15 years, expanded the house to include additional rental rooms and a shop. His father lives with his family on the adjoining plot.

Steven runs three businesses: water vending, electronics retail (goods imported from Togo and sold locally), and public showers. From the shower and water business, his profit is some $5–$10 per day, or $170 per month.

He employs two workers to manage this business, paying them $30 each per month. The electronics business is relatively new, so income from that is minimal. He also owns five rental units, from which he earns a rent of $30–$35 per month each, that is, $170 total. The total household income is $340 per month.

There are three structures on the plot: the walls of the main house are made with wood, and those of the rental block with concrete blocks; the roofs are made of tin sheets. There are seven rooms in all, two of which are occupied by Steven and his family, and part of the space in one of the rooms is used for the electronics business; the other five are rented out. Cooking is done outdoors. There is no official title to the plot or the house, but Steven says he "bought this plot of land from the so-called 'caretaker' of the land." He has a receipt for the payment, but no site plan.

There is one meter for electricity for all the households on the plot. The bill is typically about $55 per month, most of which gets covered by the renters and the business expenses; his share is about $5 per month. There is a water standpipe in the plot, which is used for the household consumption as well as for water sale. The total bill per month is $110, but it is recovered with 150 percent profit from the water sale and shower business. There is no toilet in the house. They use the privately built public toilets, for $0.10 per use, amounting to a monthly expenditure of about $35. On solid waste disposal, Steven says, "The waste container is near the market, very far from here, so we simply throw our garbage at the beachside."

Steven spends a little over half of the household income on food, about 10 percent on services (water, sanitation, electricity), and 17 percent on the children's education. Steven does not have any formal bank account, and says, "Banks are cumbersome; they require too much documentation. I tried once to open an account, but then decided against it." He has a *susu* account to which he contributes $2 daily, or $65 per month. When he needs more money, he says he borrows from friends. "Banks are very difficult to approach. One of my friends went to the bank, and waited for eight hours before he was attended to, and ultimately returned with nothing."

Florence

Florence cooks and sells local food. She lives alone in a one-room shack and has set up her cooking area just outside the house. She has three employees, each of whom she pays $1.5 per day. She still makes a profit of $6–$7 daily, which translates into $200 per month, sometimes more. Much of this she remits to her family—husband and five children—who live in her ancestral village. "I am in the city for the business, to make the money," she says. "There is no room for the kids here; they are better off attending school back home. I visit them occasionally." Florence's husband is a driver, but unemployed, so she is the only breadwinner of the seven-member family.

Florence came to the city ten years ago. She rented a room in this settlement—an informal settlement in a flood-prone area—which she shared with four other women. She set up her work base in the neighboring market doing a range of activities: selling plantain and working as a shop assistant and a porter. She did this for four years, saved some money, and finally requested a plot of land from the local leadership, which she was granted five years ago. This is where she built her house and set up her food business. "I have heard some rumors about the government wanting to move us. If that happens, I will pack up and go to my village. What's the point if I cannot continue my business? Of course, if they give me a place to set up my business, I will stay."

The house is essentially a one-room temporary structure made of wood and tin sheets. In the front is a shed, which Florence uses as the cooking area for her business. Florence has an electricity connection from her neighbor, for which she pays a flat fee of $2 per month. She buys water from the neighbor. She uses a private pay toilet, which costs $0.05 per use—about $18 per month. She spends an additional $15 or so per month for the public showers. There is no organized system for garbage collection or disposal. Trash is thrown into the lagoon by the settlement, and wastewater is disposed of in the unpaved street in front of the house.

Given the food business, Florence's expenditure on food is only about 17 percent of her income. She spends another 5 percent on services (water, sanitation, electricity) and sends the bulk of her income back home (44 percent). Florence saves about $2 per day into a *susu* account. She says she is not interested in any other schemes for saving money or accessing loans. "I have had bad experiences in the past, and don't trust them. I would much rather live within my means, save on my own, and be responsible for myself and my children."

INFORMAL SOLUTIONS TO FORMAL HOUSING

Jasmin

Jasmin is a 38-year-old living in an informal settlement with her husband, a 7-year-old daughter, and three other dependents; these include her 25-year-old sister and her infant daughter and a 20-year-old brother-in-law, who is a recent graduate but unemployed. Jasmin was in the last stage of her second pregnancy during the interview. Her husband is a driver in the public sector, earning about $120 per month. His job also takes him outside of town two to three weeks in a month, when he is able to draw additional savings from his daily subsistence allowance and purchase cheaper food items from outside of town.

They are members of a self-help cooperative that was started in 1992 with 168 households. The number of households has now increased to 198. According to a committee member of the cooperative, the cooperative had initially approached the government with a request for land and was met with conflicting responses from different segments of the government. In 2001, a court ruling approved the legalization of this cooperative, but apparently the

Box 3.5   Informal solutions to formal housing.

Jasmin is member of a housing cooperative that 'illegally' purchased and developed farmland into a residential colony. She moved to the area some 11 years ago, rented a place, and began to construct her house incrementally. The building is still incomplete, but recently, the government imposed a rule against further construction, so the work has now been put on hold. But she, like other members of the cooperative, is determined to get this area legalized and serviced, in the absence of any other formal sector housing solutions that are affordable.

Sofia is part of an informal circular group savings scheme. It was through this scheme that she has been able to construct her 2-room house.

city government intervened and appealed against it. As a result, they decided to go ahead and purchase land from farmers and establish a housing development for their members. Land, about 4,000 m² in area, was purchased from four or five farmers and divided into 198 plots of 160 m² each. Each family paid $650 for its plot and has been contributing additional sums (including a cooperative fee of $3 per month) for other services including roads and water. The settlement has a well-laid-out road network, adequate rights-of-way, and drainage channels. Several water points are spread across the settlement to service all the households. Electricity was still a problem at the time of the interview, but the committee is in negotiation with the electricity board to obtain a supply.

Jasmin is currently a renter, living in a furnished house belonging to an expatriate. Her rent is $25 per month. She was previously residing in the city center in a private rental house for five years. She moved here two years ago to oversee the incremental construction of her house in the adjacent plot (started eleven years ago). Unfortunately, just as the construction of her house progressed to the plinth level, the government imposed a stringent ruling against any new construction. She is now awaiting the much anticipated "legalization" so she can resume construction.

Water is obtained from a shared water point, at a cost of $0.10 for 25 liters; the family spends approximately $3.50 a month on water. Electricity is drawn from a neighbor from another cooperative, and given the relatively large number of electrical gadgets (fridge, computer, TV/VCR, oven, etc.), they also pay a higher electricity bill in the range of $18 per month. The house has its own pit latrine. The family hires a private garbage collector and pays $1–$1.5 a month for his services.

Jasmin, although much better off than most of her neighbors, is pressed with the decision of moving back into the city in the event the government does not formalize this settlement. She has already spent a substantial sum on the construction, and estimates an additional expenditure of $4,500–$5,000 for completion. Since the bank will not loan them the money and the local microcredit facility only offers very small loans, she says her sister-in-law in Israel will finance their house construction. Although she has already paid the cooperative about $600 for the land, she is willing to pay the government another $600 for a land title, so long as the monthly payments do not exceed her present level of saving of roughly $20–$25 per month.

Sofia

Sofia is a 56-year-old primary-school dropout who works as a farmer. She has three daughters, two of whom are married and living elsewhere. She lives here with her husband, a 30-year-old daughter, and two grandchildren (15 and 12 years old, from the eldest daughter's first marriage). Sofia has access to about 0.4 hectares of farmland near her house on the slopes, where she plants rice and vegetables; much of the crop yield is used for her family's consumption rather than sale. So, essentially, she has no real income. Her 64-year-old husband is a salaried employee of a private firm and gets $50 per month plus benefits.

Sofia migrated to this city from her village hometown in 1973 with her husband. They lived in a single rented room in another settlement for six years. In 1979, they moved to another area in the city where they spent the next 18 years, before coming to this settlement. By this time, they had three grown-up children. They initially lived in rented accommodation, but within a few months they were able to find a piece of land that their neighbor was interested in selling. They bought this plot for $170: "It was relatively cheap," says Sofia, "because it is on the slopes where we are not supposed to build. Still, today this plot will cost about $500."

Sofia's house is a two-room structure, which Sofia and her husband built incrementally over three years, using self-made mud blocks. Cooking and washing is done outdoors in a small yard in front of the house. She is not aware of the title or license the government has recently been encouraging residents to acquire, and says they have not started paying property tax just yet. "The sub-ward officials conducted a mapping exercise, in which they numbered all the houses in the area; ours was not numbered, so I don't know what the status is," she says.

The house has no electricity. For cooking, they use twigs of wood collected from the area and kerosene, which costs about $4 per month. Water is purchased from a neighbor who owns a well. The charge is about $0.02 per bucket of 20 liters. They spend about $2 per month for water. There is a pit in the plot along with a bath area. Water from the bath is channeled down the slope into the river below. Garbage is disposed of on the hill slopes on the side of the house.

The two grandchildren attend public primary school, about three miles from the house; the children walk there. Tuition is free for the government school, but the children's allowance for transport, plus their uniforms and books, and so on, costs some $7 per month on average. For medical treatment, the family uses a private health clinic nearby. Like many others living here, Sofia says malaria is very common.

The bulk of the household income, 42 percent, goes on food, 8 percent on fuel, 3 percent on water, and another 13 percent on education and related expenses. Sofia is a member of an informal circular group-savings scheme, where a group of women get together and contribute a certain amount every month. In fact, it was through her participation in this scheme that she was able to finance the construction of the house, she says. She would contribute $4 per month in a group of 16 women, and was able to raise $64 when it was her turn to collect the money. Today, she has an account in the same sort of scheme where she contributes $8 twice a month. "This works well for me," she says. "I don't participate in any other savings program, and am not interested either. The same applies to loans: I don't like to borrow."

Box 3.6   Informal settlements as places of refuge.

Thandi was once married to a famous rock star (in the photo frame), and came from a well-to-do family. Today, she is unemployed, alone, and an alcoholic, and lives in a slum, renting one room in a wooden structure and shares a toilet with some 50 other renters on the plot.

David is a 38-year old high school graduate. Originally from a middle income family, he faced a series of unfortunate life-changing events, and was convicted of a crime. Today he is unemployed, an alcoholic, and lives in a run-down wooden shack with no toilet or clean running water. Regarding prospects for productive employment he says, "They are estate people; we are ghetto people. They will only help and employ their own . . . but I will succeed the day I have the opportunity."

INFORMAL SETTLEMENTS AS PLACES OF REFUGE

Thandi

Thandi is a 50-year-old unemployed single woman. She came to the city with her husband at the age of thirty. Her husband, now deceased, was a famous singer. "Everyone in the country knows him," she claims. However, after a few years of marriage, when Thandi could not bear a child, he left her for another woman. Thandi then moved in with her girlfriend in another settlement who had a small alcohol-brewing business. She stayed with her for a few months while assisting her in the business. With no education to get a "decent" job and the need to reestablish herself as a single woman, Thandi borrowed $70 from her friend and started her own alcohol-brewing business. She then moved into an independent unit—a single rental room, with a toilet and water connection, for a rent of $8 per month—which she says was "much better than this house." The business was a big success in the first few years. She fetched an average profit of $7–$8 every two to three days, or about $80–$100 per month.

When her father died, she used all her savings towards building and furnishing a new house for her mother in the village. "I didn't have any children, so I did not need the money at the time," she says. "Besides, my business was doing very well, and I thought I could easily earn enough to support myself." Unfortunately, three years after she had started her business, the prospects began to dwindle. She closed that shop and moved to another settlement hoping to reestablish her business there. After three months of trying, she realized it was no use. "The police controls had become stronger, and the clientele had diminished." So she moved once again, to this current accommodation, which was cheaper. Attempts to restart her business here were again in vain. With no alternative livelihood and no work to do, she started visiting the pubs in the area. "Liquor is cheap there, cheaper than any other commodity," she says, "only $0.14 for a pint." Over the years, she made other friends who frequent the bars. Now some invite her and pay for her drinks.

For food and rent money, she relies on her 36-year-old brother, who does part-time jobs as a metal worker. He gives her about $10 every month. Thandi showed photos of her family—her pop-star husband (now deceased) and the rest of the family members. The pictures were neatly framed in a carved wooden frame. They are proof of the "decent" life she enjoyed in a relatively well-to-do family, until life took a downturn. This is also evident from the way she has kept her unit. Despite the dilapidated structure she lives in, the interior of her house is neat and clean: the tin walls are lined with cardboard sheets; water containers are stacked in one corner, the cooking stove and utensils in another; a curtain partitions the bed from the seating space, comprising a few stools and a table. The floor, although unpaved, is clean.

Thandi currently rents one room (approximately 3 m × 3.5 m) in a barrack-like block of rooms, constructed with wood and iron sheets (walls and roof). Cooking is done indoors in a small corner of the room. There are two toilets on the plot, shared by 17 households (about 50–60 people). However, there is no lock on the doors, so passersby also commonly use the facilities. When the pit is full, the landlord arranges to get it emptied. However, during the time of the interview, the pits had reportedly been full for a week. The people were either continuing to use them despite that, or other public toilets, or the bush. Thandi buys roughly one jerry can of water (20 liters) per day for drinking and cooking, at the price of about $0.07 per jerrycan. The house does not have an electricity connection. Garbage is disposed of in the street, and there is no drainage system within or outside the plot.

David

David is a 38-year-old alcoholic living in an informal settlement alongside a river creek. David is a high school graduate from a private school, and speaks fluent English. He is fond of reading and has a small collection of biblical literature. He does some part-time work, including assisting women with farming twice a week, and brewing local liquor. He earns less than $1/day from farming work, and another $1 or so from the brewing job. Previously, he would also do scavenging at a nearby dumpsite, but not any more.

David currently rents a single room (approximately 2.5 m × 3 m) in a barrack-like structure of five rooms, constructed with wood and corrugated iron sheets (walls and roof). The floor is unpaved. There is basically no furniture apart from a worn-down mattress and some clothes are swung on a rope stretched across the room. The rent is $6 per month. An outdoor cooking space is shared by the five rental units on the plot. Plastic sheets are used as fuel for cooking. There is no water, electricity, or toilet on the plot.

David came to this settlement as a young man in 1989 as part of a government resettlement program with his mother and three brothers (two older and one younger). His father left home for another woman when David was a young child, but he saw him occasionally as he was growing up. They were a well-off family. One year after moving here, in 1990, David's eldest brother was stabbed to death. In 1995, his mother died of a heart attack. David sold the house that belonged to his mother for $85. Half the money was used for his mother's funeral, and the other half toward his youngest brother's education. He then got himself a rental unit from his mother's friend for $5 per month, where he stayed for three years, and then to another place for two years until he was evicted because he could not pay the rent anymore. He then came to this house and has been here for three years.

With limited skills to get a regular job, and few opportunities, David learned how to make the local brew (illicit liquor) from his friends. That got him part-time work in the settlement—making brew for a local businessman for which he got paid $1/day and received free liquor on the side. This combination of "idleness, lack of work, and plenty of cheap 'wine' gradually turned me into an alcoholic," he says. Concerned that this environment would ruin his youngest brother's future, David sent him to the home of his father's second wife for schooling. David continued to stay here with his other brother, and they worked together in the brew business. In 1997, David was arrested, along with his friends, under suspicion of theft. He denies stealing but admits being involved in selling the stolen items. He was in prison for four years, where he undertook a short six-month training course in carpentry. After his release, he wanted to start his own carpentry business, but did not have the money or adequate skills to get started. He needed more training but did not have resources for that either. So he went back to making the liquor.

In 2004, David's other brother contracted cerebral malaria, impairing his nervous system. Unaware of the seriousness of the illness, and unable to afford treatment, his condition went largely untreated, until he committed suicide late last year. David claims that most of the youth of the settlement are in a similar situation with respect to jobs and social life. "Marijuana is the most common drug among the youth: cheap and widely available. One stick of marijuana can be purchased for about $0.15. It's everywhere, and provides a cheap source of entertainment," he says. "And prostitution in the settlement is directly linked to poverty and alcohol." David himself was a marijuana addict for ten years after he graduated from secondary school. "No more," he claims. He came clean after he was diagnosed with TB and underwent an extensive three-year treatment for the infection. Regarding prospects of a job outside of this settlement, he says, "They are 'estate' people; we are 'ghetto' people. They will only help and employ their own . . . But I will succeed the day I have the opportunity."

4   Policy and Practice
The Missing Link

This chapter presents, the disconnect between what people want and need versus what planners decide is "good" for them is a very fundamental problem in development—the missing link between planning and practice—that might be exacerbating the problems associated with slums and informality rather than addressing them. We illustrate, through qualitative research and case studies, the mismatch between the needs and expectations of the residents of informal settlements and the solutions that governments, development agencies, and planners propose for them. We show that despite the lip service paid to consultation, participation, and community involvement, decisions are often taken without considering the wishes of affected communities.

LESSONS NOT LEARNED

As mentioned in the earlier chapters, informal settlements are usually seen as unwelcome "dens of crime" and "eyesores" that "destabilize" modern society. In the previous two chapters we tried to give them a human face, and to demonstrate that, in spite of physical and economic barriers, so-called slums are a productive segment of society that consume the least while contributing substantially to economic growth. From the macrolevel and from a historical perspective, informality is the first stage in a natural progression toward urban growth and development: after all, many of the world's largest cities have a history of slums and informal development.

> Yesterday's slums are today's world class cities. Britain is not the only industrial country to suffer from slums and wide intracity divisions in welfare during the earlier phases of development and rapid urbanization. . . . Indeed for "world" cities such as London, New York, Paris, Singapore, and Tokyo, slums can, with the benefit of hindsight, be viewed as part of their "growing pains."[1]

This transformative effect of urban development is more than a physical one: the strength of urban economies and the opportunities that cities offer provide people an opportunity to flourish: A metropolitan economy, if it's working well, is constantly transforming many poor people into middle-class people, many greenhorns into competent citizens. Cities don't lure the middle class. They create it.[2]

That said, this progressive thinking has been slow to take shape. A major shortcoming has been the inability—and even apathy—of the powers that be to really understand the situation. As planners, we are trained that in order to propose the right solution, one must understand the problem. This requires us to carry out accurate assessments. From there we are supposed to propose feasible solutions and provide adequate tools for implementation. Unfortunately, however, the assessment phase is often cut short in favor of "projects" that are essentially photo opportunities and keep donor money flowing to meet annual investment and expenditure targets. The shortcutting of the assessment phase and an over-reliance on secondary data—which is often not verified or verifiable, or dated, or simply inaccurate—commonly results in misrepresentation of the real problems. This in turn leads to oversimplified solutions influenced by overrated "best practices" and out-of-context buzz words. For example, proposing physical upgrading when the real need might be one of skills development and employment generation, or proposing large investment of scarce government resources on subsidized housing programs for the "poor" which are, on one hand, ill-suited and unaffordable to the beneficiaries, and on the other, financially unviable for the government. An interesting example in this regard is the continued adoption of the failed public housing model of the 1960s and 70s by developing-country governments as the overarching solution to slums, informality, homelessness, and poverty.

Slums and poverty are inextricably linked. The growth of slums and informal solutions to housing in recent years has resulted from a number of social and economic factors, including urbanization, declining wages in relation to inflation, growth in single households, increase in demand for affordable housing, and the inability of the formal sector to meet the demand with affordable solutions. But in addressing the issue of housing, a common tendency has been to emphasize its physical aspect, that is, the structure and infrastructure, without adequate linkages to the social and economic development aspects—livelihood/ skills development and training, microcredit facilities, savings and income generating opportunities, health and education facilities, for example—which are critical for basic human development. Often, as a result, neither are the people able to afford the "solution" imposed on them—whether housing or infrastructure improvements—nor does the government recover costs, making such efforts unsustainable in the long run.

ONE SIZE FITS ALL

Indeed, history repeats itself. We need look no further than the public housing projects in United States to understand how such programs virtually destroyed the social fabric of entire communities, turning them into "bad" neighborhoods ridden with poverty, crime, drug abuse, and broken families. The projects obviously attempted to provide decent housing to the poor, but in the absence of adequate access to educational or health facilities, or employment opportunities, they essentially locked them into the vicious cycle of poverty. This past decade has seen the demise of this public-housing approach in the United States; federal dollars are now being used to tear down and revitalize many of these neighborhoods to make them more "livable."[3]

The irony in this case, however, is the time lag between what the United States has learned from decades of experience and its transfer to developing countries, many of which still base their housing programs on the rejected U.S. model. And despite the very distinct and disparate conditions with regard to land and tenure issues, socioeconomic conditions, and physical infrastructure—by region, by country, and by settlement—the "solutions" bear a frightening resemblance. Whether on the streets of Mumbai, in the slums of Manila, in the chiefdoms Swaziland, the tenements of Nairobi, the underserviced *kebeles* of Ethiopia or Eritrea, or in the small towns of China, a common denominator is medium-rise public housing[4] as a means to "densify prime land" and/or to "clean up" the city. This one-size-fits-all approach has little bearing on the real needs or wishes of the target group itself (see Boxes 4.1, 4.2, and 4.3).

MYTHS AND MISCONCEPTIONS

This section illustrates how the local situation is so easily misconceived in the absence of adequate field research and engagement with affected communities, as a result of which many of the "pro-poor" projects in effect become "anti-poor." Myths and misconceptions about informal settlements, about what they are, what they need, and so on, are abundant. Here we use examples from our experiences to illustrate the fallacy underlying some of these preconceived notions. These will not come as anything new to many readers who are familiar with this type of work. However, as our experience shows, in reality there are many who are not, and hence these issues are worth discussing.

Box 4.1 Housing solutions in Addis Ababa, Ethiopia.

In 2004, the Addis Ababa City Council committed about half of its annual capital budget towards the construction of housing. To quote the (now ex-) Mayor of Addis Ababa from the Fortune (Addis Ababa, Ethiopia, July 18, 2004):

> *Old houses will be demolished and new houses will be built in selected pockets to embellish the appearance of the city. They will be replaced by apartments of three and four storeys that could fit the status of Addis Ababa.*

The pilot housing project, designed and implemented by a development agency, was completed in 2004. It comprises some 700 units, including studio, one-, two-, and three-bedroom units, promoted as "low-cost" housing. While taking care not to confuse "low-cost" with "low-income" housing, this government-funded initiative immediately raises a red flag: the housing is supposed to be for the "poor", but not even the one-bedroom units are affordable to the bottom 65 income-percentile of the population. This begs the question: who does the project really benefit?

An aerial view of a kebele settlement in Addis

Government subsidized "low-cost"—but not "low-income"—housing project

Source: Field research by Ashna Mathema in 2004.

Myth: Slums are inconsistent with "development" and an inefficient use of land in cities.

Fact: Slums are a symptom and direct outcome of bad housing policy or dysfunctional land and housing markets, and not the other way around.

In this regard, we categorically say that slums and informal settlements are among the more efficient and organic forms of development in terms of spatial layout and densities at the time of their formation. As land values have increased, however, these areas are deemed by planners and policymakers as an inefficient use of prime land, which makes them an attractive target for other uses.

Box 4.2   Medium-rise buildings in Metro Manila, Philippines.

In the early 2000s, the government of the Philippines was a strong proponent of five-story public housing, better known locally as medium-rise buildings (MRBs). These MRBs were viewed as a solution for resettling much of Metro Manila's 4-million strong squatter population.

What became apparent, however, was that this type of housing is inappropriate not only in terms of the physical spatial needs of the target population, but also potentially disastrous from the social standpoint, and in terms of affordability and cost recovery, for several reasons: one, the average size of the units was 20 m², including a bath and kitchen; two, the MRBs virtually eliminated any potential for secondary sources of income, such as shops; three, their physical layout was unsuited for community interaction, negatively impacting the critical social safety nets; four, maintenance was a major problem, with some roofs developing leaks less than six months after handover; five, cost recovery was a problem, since there was no real understanding of either affordability or the willingness of the 'beneficiaries' to pay for these units.

Despite all of these problems, the MRB continued to be a predominant aspect of the government's housing strategy.

Stilted housing in a slum along a creek in Manila.   A medium-rise building where slum dwellers were resettled.

Source: Field research by Ashna Mathema for ADB's *Metro Manila Urban Services for the Poor* project, PADCO (2002).

An often overlooked reality, however, is that informal settlements, particularly those in large cities, are in fact very high density and typically utilize both built-up and open areas to their maximum potential. In the most fundamental sense of spatial development, some might even be comparable with "smart" development, with a rich mix of land uses, located close to transportation corridors and job centers. This is not to say that inefficiencies don't exist, but rather that they mostly come from external factors—housing policy, certainly with respect to land titles, the poor delivery of public services, and so on—over which these communities exercise little control.

Box 4.3   Resettling farmers in China.

As towns grow, they need land for expansion. Whether this land is for industrial development, or for relocating existing land-uses to facilitate efficient planning in the city center, in China it is imposing pressure on local governments to acquire agricultural land from the farmers and convert it to other uses. The resulting displacement of resident households impacts two population sub-groups: one, farmers whose land is being acquired, and two, families that are being resettled to the peripheries of the towns to allow for more productive and profitable use of land in the town center. This is the type of housing they are being given.

A local points to the plan of an on-going new development.

A government housing resettlement project for farmers whose farmland was acquired for industrial development in Wutong, Shanxi.

Source: Field research by Ashna Mathema in 2005.

Consider, for example, Addis Ababa, Ethiopia. Unlike other developing countries where informality is typically associated with some notion of illegality, in Addis, it is much of the legal housing stock that is "informal." As discussed in Chapter 2, this is the government-owned *kebele* housing, essentially single-story rental housing, occupying vast tracts of land in prime locations. Many residents are middle-income wage earners. The government calls these "low-rise, low-density" settlements which are "an inefficient use of prime land." The plan is to demolish these structures and replace them with "modern" medium- and high-rise housing to improve the image of the city.

On probing a little deeper, we realize that it is not the residents who can be blamed for the poor quality of the structures they occupy; the government owns the houses, and the renters are not permitted to make improvements. Rents in the *kebele* housing have remained nearly constant for the past 20 years, and are rarely collected. As a result, maintenance is extremely poor, and basic services non-existent. But despite the poor services, people choose to live here because it is often the only affordable solution. And there is very little turnover: once occupied, the renter will keep it for as long as possible.

Box 4.4   "Formal" informality: *Kebele* housing in Addis Ababa, Ethiopia.

Rents from *kebele* houses are very low—they have not been adjusted since the 1970s—and not even enough to cover the maintenance of the houses, let alone serve as a source of revenue for the city. The *kebele* councils, for fear of heightening compensation liabilities, do not permit the residents to make substantial renovations to the houses. "Allowable" changes include only basic maintenance of the structure, or upgrading the building material; the floor plan of the original house must be adhered to. Occasionally exceptions are made.

Similarly, new construction is not permitted on an empty lot or abandoned house, except for building communal toilets and kitchens. The implication is that even if a plot has additional space to build an additional room, or add a personal toilet or a kitchen, this is not permitted by law. Interviews with some CBOs working on housing and/or maintenance revealed that even though they had generated sufficient funds (from community contributions and NGO/ donor support), they had been unable to carry out any substantial improvements to the houses, due to the legal constraints. This is despite the fact that the money to rebuild would come from the community funds, not the government, and that the house would still, by all legal standards, belong to the government. Occupants, on their part, do not engage in improvements of their homes partly due to lack of incentive or resources, but mostly for fear that rents would be increased if the houses were improved.

A house where a loft space has been created by the renter so as to create additional sleeping room. This is about a meter in height, and accessed by a ladder through a hole in the floorboard below.

Source: Ashna Mathema, *Housing in Addis: Background Paper*, 2005 [Unpublished background research for The World Bank, funded by the Danish Trust Fund].

To get around the restriction on improvements or expansion, some people have built mezzanines or loft spaces above their one-room structures to generate additional sleeping space (see Box 4.4). As a result, there are often about ten people sharing on room of 10–12 m^2, which makes them anything but "low-density." The problem therefore underlies the stringent policy that bars these households from making improvements, and at the same time, holds no one responsible to maintain the housing stock.

Similarly, in Lagos building codes and regulations have had the unintended effect of constraining the already constrained supply of land, and thus marginalizing about 70 percent of the city's population that lives in slums and informal settlements. Land, in general, is a scarce resource, but its limited availability is even more pronounced in island cities, or those that are confined geographically. Many such cities thus try to maximize land use through larger floor-area ratios, higher densities, and so on. The resulting urban form is a manifestation of the demand-supply balance, which is why cities like New York, San Francisco, Hong Kong, and Singapore—all coastal cities—are well-known for their skyscrapers and high-density developments, while cities located in flat plains without major water barriers—like Paris, London, and Berlin—are not. Not all cities, however, respond effectively to geographical and topographical constraints. Lagos is like many other cities in developing countries which, rather than increasing floor space by expanding vertically, attempts instead to curb the demand through instruments such as ceilings on urban land ownership, or constrain the effects of demand on prices by placing ceilings on rents, land development, and land use. In reality these policies, however, often have an effect diametrically opposite to what they set out to do, which is to increase access and affordability; instead, they end up artificially constraining supply and increasing prices (see Box 4.5).

Most of the housing that is currently being built in Lagos by the private sector consists of 3- to 4-bedroom bungalows and flats in luxury estates costing, on the conservative side, between N15 million ($150,000) and N25 million ($250,000), which is beyond the reach of the vast majority of the population. There are also cheaper 2- to 3-bedroom units being built in the outer suburbs of the metropolitan area that cost between N1 million ($10,000) and N5 million ($50,000) which are more affordable, but those are typically very inconveniently located and poorly connected by public transit.

Similar plot size regulations are prevalent in many other African countries. In Ghana, the minimum plot subdivision is 450 m². Tanzania's minimum plot size per building regulations is 400 m², the inappropriateness of which is evident from the fact that 70 percent of Dar es Salaam's population lives in "informal" settlements with plots averaging about 150–200 m².

Similarly, in Mumbai, the development pattern has been drastically restricted by the building height restrictions embodied in floor space index (FSI) regulations. The overall FSI for the city is one-fifth to one-tenth of the level of other large cities, affecting how people locate and the types of commuting patterns they follow. But Mumbai regulates

Box 4.5   Subdivision regulations promoting slums and sprawl in Lagos.

The minimum plot size for residential developments in Lagos ranges from 1000 m² in areas zoned as "low-density" to 648 m² in "high-density" areas.

Density	Min plot size (m²)	No. of dwelling units allowed on plot
Low	1000	3
Medium	864	7.8
High	648	17.5

This requirement is excessive by any standards, and is a critical impediment to reducing, incrementally over time, the level of informality in the housing sector. There are several reasons why this is the case, the first and most obvious being that a large plot requires high purchasing capacity, which makes it beyond the reach of poor households. These prices can be gleaned from a local real estate newspaper, where plots average 1,000–1,500 m² in area and cost millions of naira, even tens of millions, depending on the location. Second, unless countered with regulations that strongly discourage low density development, this sort of subdivision regulation contributes to inefficient urban sprawl that is a challenge for transport, air pollution, and infrastructure provision in general, with high costs of service provision per capita, exacerbating the already overwhelming demands on service providers.

Senior staff in the Lagos Urban Development and Physical Planning Ministry explain the rationale for this sprawl-promoting land use policy as follows: "... *we Nigerians have an attachment to land, we like a lot of space.*" While supporting the consumption of "a lot of space" may be a well-intentioned aspiration, the fact is that it benefits only the top 5 percent of the population, which may not be the intention of the leadership in Lagos to begin with. Its unintended impact, however, is creating severe living space constraints for the citizens of Lagos—not just the poor, who constitute almost half the population, but also the middle class.

Source: Ashna Mathema, *Slums and Sprawl: The unintended consequences of well-intended regulation*, Background Paper (Unpublished research for World Bank), 2007.

much more than building heights; it also allows some of the city's most valuable land to sit idle, effectively wasting the imputed "income" that could have been realized from the use of this property.[5] It comes as no surprise, then, that more than half of Mumbai population lives in slums.

Myth: The poor live in subhuman conditions in slums and informal settlements, with no potential for economic development.

Fact: People make rational economic decisions regarding their choice of housing.

While slums house the vast majority of the poor, it is well known that it is not only the poor who live in informal settlements. Middle- and lower-middle-income families often live in these areas, because of their inability to access any other formal sector housing that is affordable. Box 4.6 shows how, in many cases, however "shabby" the houses are from the outside, they are fairly well-decorated and well-equipped from the inside. Planners and policymakers are quick to label informal settlements as "inadequate and substandard" because of their external appearance that does not fit within the neatness of building codes, but often the interior is well organized and equipped with a surprisingly wide range of assets. This was very evident in the Philippines, where practically every household who was interviewed, no matter what the house was made of, owned a boom box, and had relatively well-organized interiors.

An often overlooked fact that it is just the essence of "informality" that lends people the opportunity for economic development. This is illustrated by an example from a slum in Nairobi (discussed in Chapter 2), whose physical appearance leaves little doubt about the squalid living

Box 4.6   Looks are deceptive: Metro Manila's "slums."

The photo on the left shows the exterior of a house in an informal settlement. The top-right photo is the interior of the same house: a poor quality exterior does not mean a poor quality interior. The bottom right photo is an interior shot in another similar house.

Source: Field research by Ashna Mathema in 2002.

conditions and poor level of services in the settlement. However, a more in-depth analysis reveals that this particular settlement is among the more entrepreneurial informal areas in the city. A case worth mentioning is that of Desta, a 57-year-old woman entrepreneur, who came to the city twenty-four years ago with KSh140 (approximately $2), four children, and no husband. Today, she runs five businesses simultaneously, which she established incrementally over the years, and has an income of KSh60,000 ($850) per month (see Box 4.7). She could easily afford a formal house, she says, but all of her businesses are *in* the settlement, and she would rather spend the bulk of her money on her children's higher education instead of on a "fancy house." Both her daughters are currently university students. Desta's is a story of conviction and success, perhaps an exception. Still, the fact remains that people often make a rational choice to live in these areas for the many advantages they offer.

> *Myth: People in informal settlements have no motivation or capability to improve their living conditions, which has negative externalities on the rest of the city residents; hence, government must play a major role in providing them with free or subsidized homes.*
>
> *Fact: Marginalized communities are perfectly capable of leading and planning "proper" developments; they just need a system that facilitates their drive to get decent housing for themselves.*

Addis Ababa illustrates this well, with its increasing incidence of so-called illegal developments. People are coming together in groups called cooperatives, illegally purchasing land from farmers on the city periphery, and essentially "developing" the area. They are careful to meet all the prevailing planning standards—in terms of minimum plot size, road width, right-of-way, and so on—in the hope that they will be legalized sooner or later (see Box 2.6 in Chapter 2). With each household contributing a fixed sum every month toward capital and maintenance costs, just as in any formal housing association or cooperative, many of these cooperatives have managed to get the basic trunk infrastructure for electricity and water supply.

The quality of houses built on these plots varies, depending on individual affordability, but in general, all permanent construction is in compliance with city regulations. In other words, these are "planned" developments, which people have undertaken in the face of government inaction to address the distorted land and housing markets, and the extreme short supply of affordable housing. Still, they are deemed "illegal" and constantly under threat of demolition from the inspectors.

In Accra, many of the poorer families occupy traditional "family houses," with densities averaging four persons to a room. In some cases, 15–20 persons

Box 4.7   House in a slum: An economic choice for Desta.

Desta came to the city from her village in 1981. She had just divorced her husband. With 4 children, and no income, she came here in search of work. First, she first stayed with a friend in—a neighbor from the village—who was living here. She started a small vegetable vending business. She would walk 2 hours back and forth to the city centre daily to sell vegetables. After 4 months, she was allocated a plot of land by the leadership. She built a house out of packing cartons and plastic sheets, where she lived with her 4 children for 6 years. Her vegetable business slowly picked up, and she reinvested the savings into other small businesses—selling cattle and chicken.

In 1987, she requested the leadership for a larger plot to accommodate her growing children, which she was granted. Here she built a larger house with a total of 8 rooms of which 3 were used by the family and 5 were rented out. With money from rent, she started a water vending business. At the same time, she built another 6 rental rooms on the first plot. By 1996, her 2 sons were married, so she requested another plot for them from the Chairman. For this she paid about $30. She now lives in this plot, and has given her sons a room each in the second plot. The first plot is now houses a shed for cattle /chicken.

Just as she was able to reinvest her income from rent into building new houses, she also reinvested the profits from the businesses to start new ventures. Today, she runs/manages five successful businesses: (i) selling wooden poles, from which she earns about $200-250 per month; (ii) selling water, which brings a monthly profit of $350 per month: (iii) selling goats/ chicken; (iv) pay-to-use toilets, which generate an income of about $100 per month; and (v) rental rooms, which get her a net profit of $100 per month.

Desta's total income from all of these business ventures together is in the range of **$850 per month**. This is a substantial amount, yet she chooses to live in a 2-room house built of iron sheets and wood (plus a rudimentary toilet and bath area). She could easily afford a formal house, she says, but all of her businesses are in the settlement, and she would rather spend the bulk of her money on her children's education instead of on housing: "I want my children to benefit from the fruits of my labor." Her older daughter is enrolled in a Masters program in business management, and the younger in a private pharmacy college.

Desta outside her house made of tin sheets and wood

TV, radio, cell phone, other gadgets in the house

Source: Field research by Ashna Mathema in 2005.

share a room: the room is essentially used for storing the belongings, and the people sleep outside. Families often split up in different houses (ancestral homes) due to lack of space: husbands live separately from their wives, the children split up, or are handed over to the in-laws or grandparents to look after. Despite all of the inherent problems with "family houses"—poverty, multiple ownership, and lack of registered titles—the culture of cross-family support lends itself as a social safety mechanism. The poorest in the family at least have a roof over their heads, however modest and/or inadequate. Jamestown is one such settlement which is extremely dense and very vibrant, during day and night alike. Every square inch of the land is used to its maximum potential. Chapter 3 presented several examples of individuals and families living in informal settlements whose survival strategy it is to work hard and be the most productive using minimal resources.

SUCCESS STORIES THAT ARE NOT

Success stories and so-called *best practices* are critical from the perspective of learning what to do and how to do it without reinventing the wheel or repeating the mistakes of the past. However, success stories also have a high potential for photo opportunities and run the risk of being overstated, for example, by manipulating data to make the project look good, or simply being based on false assumptions resulting from lack of monitoring or follow-up of what really happened. And the resulting "best practices" are often not "best." Two examples from Kenya illustrate this.

The first is about a recent policy reform that took place in Kenya, which introduced free public-sector primary-school education. Two years after the reform was enacted, it was found that enrollment rates rose to over 90 percent, which suggested considerable progress from prior years. Both government and donors were quick to take credit, and rightly so. To get such results in such a short period was remarkable. However, there is more to this story than official data suggest, pertaining to the policy's impact on the quality—rather than the quantity—of education.

A qualitative field study conducted by Mathema[6]—largely, household interviews—suggested something remarkably different. They revealed that the quality of education in public schools had in fact deteriorated significantly since the introduction of the reform. While the "free" system encouraged new enrollment (of children who were not previously enrolled), it also spurred transfer of students from private to public schools. This sudden increase in demand for public-school education, linked with no significant increase in funding for schools, resulted in overcrowded classrooms (with a student-teacher ratio ranging between 1:70 and 1:150), and overworked teachers who were mostly underqualified and underpaid. In the meanwhile, many of the private schools went out of business. Many parents, within two years of buying into the government program, wanted to reverse their decision and move their children back into private schools, but there weren't any to send them to. So instead, they had to incur extra costs for tuition, which were often more than what they were originally paying at the private schools. "We hardly save anything," one parent said, "because public school education is useless unless we get the children additional private tuition." In other words, while the initiative was well-intended, it ended up doing more harm rather than good, especially to the poorer families because of little or no follow-up and monitoring.

Another example from Kabul, Afghanistan, highlights the dangers underlying the importance given to superficial indicators as determinants of success. This story goes back to 2003, a year after the U.S. invasion in Afghanistan. As money was being pumped into the country, donors across the world were anxious to know where the money was going. Conferences and seminars were held, and facts and figures were presented to highlight how much progress was being made in terms of the recovery and rehabilitation of the Afghanis. One of the indicators highlighted over and over again was the increase in primary school enrollment, particularly for young Afghani girls. And while we don't dismiss those claims—after all, moving from zero up even a few notches is often more critical and more difficult than moving higher in an already established system—we contend that these numbers were being presented without a context.

Field interviews conducted by Mathema[7] in the informal settlements of Kabul revealed that most young teenage girls were *not* attending school. They had missed the boat, so to speak, and as some parents said, it was too late for them to join school at this age. So instead, they were sitting at home helping their mothers or looking after their younger siblings. But practically no event trumpeting the success of donor funds or highlighting the challenges of Kabul made even the slightest mention of this. What was being done to address the late starters for their schooling? And why were they not being discussed alongside the "successes" of the education reform as a major challenge that still needed to be addressed?

We raise this not to discredit the success achieved in enrolling more primary school-age children into school—that in itself is a difficult and laudable task—but in highlighting how the rhetoric and politicization might have inadvertently resulted in sidelining a whole generation of young girls, undermining their chance to attend mainstream school.

BUZZWORDS AND FADS

Jargon and buzzwords in the planning world are common, often used loosely to fit a specific need. A basic problem is that it is difficult to clearly define some of the terms used, as definitions vary with the context, and hence their usage. The more critical problem, however, lies in the misuse of these words, two examples of which are presented here: decentralization and community participation.

"Decentralization." In the realm of governance, the more popular buzzwords include decentralization, democratization, and anticorruption. Decentralization has its merit of local empowerment, no doubt; the problem, however, is that it is often limited to just that, that is, devolution of powers, and divorced from the critical need for simultaneous (technical and financial) capacity building at the local level.

This is exemplified by a case in the Philippines which was commended widely for its decentralization process which was initiated in 1991. As part of this decentralization, the responsibility of implementing housing projects was devolved to the local governments. Along with this came the responsibility of addressing the squatting "problem" in their jurisdictions, under a very real threat or legal sanctions for inaction. As a result, cities like San Fernando (La Union), with a mere 8 percent squatter population in 1999, began prioritizing work in the housing sector. Without the required technical or administrative capacity in housing issues, they were reliant on national government agencies that, in turn, emphasized resettlement and public housing as the overarching solution to the "housing crisis." Box 4.8 illustrates a housing project proposed by the (now-defunct) National Housing Agency to the City of San Fernando to resettle squatters,

Box 4.8   Housing for the "poor" in the Philippines.

Mid-rise Housing Proposal
Source: HUDCC

This is a rendering of a government housing proposal for resettling a small fishing (squatter community), San Fernando, La Union.

Source: Field research by Mathema in 1999.

which was highly inappropriate for the target group, a small coastal fishing community.[8]

Another case is that of Addis Ababa, where the local government under the newly decentralized governance structure is undertaking what we call urban renewal (slum clearance and resettlement) projects under the pretext of "upgrading." The government of Ethiopia is among the few countries in the world with self-funded upgrading programs and initiatives, even if the success of these programs leaves much to be debated. In the decentralized system, the lowest level of government is required to develop what are known as eco-city plans for development at the neighborhood/ward level. The premise is to "develop" prime land in the city through upgrading *and* urban renewal. Field research in 2004 revealed that much of what was happening was demolition and clearance of existing "slum" areas to make way for more lucrative developments, and moving the squatters or "illegals" into multistory housing. However, since most of this was on-site resettlement, it was called "upgrading." In other words, what they had been presenting and selling to government and donor agencies alike—with very little oversight, insufficient technical capacity, and too much decision-making power—was a product very different from "upgrading" and very controversial, given the well-documented negative repercussions of urban renewal more generally.[9]

A slightly different case is made in the third example, of China, where decentralization has resulted in distorted incentives in the public sector's role in the delivery of basic services. Decentralization in China has transferred responsibilities for financing and delivering social services to lower levels of government. Budgetary allocations from higher to lower levels of government are based on the revenue contributions from lower to higher levels of government. In other words, the poorer towns that pay less also receive very little from their county government: this implies that poorer areas will always have the least resources to address local development issues. With limited resources—technical and financial—the local governments are focusing on activities that generate most revenue. There has been a steady marketization of public services as a result of which social services are lagging far behind, particularly in small towns and villages, despite which decentralization in the country is being largely applauded.

Along the same lines, government-owned farmland in China is being continuously transformed to other uses and sold to the private sector to generate more revenue, which often comes at a high cost to small, often aging, farmers who have little say in the compensation that is decided and few avenues to cushion their financial situation.

This is not to say that decentralization is not a good thing, but rather that the process underlying it is extremely important. It must be phased in such a manner that the devolution of responsibilities happens at par with the building of technical and financial resources, so that the incentive structures do not get distorted. And while academic literature on decentralization discusses this extensively, in practice, much of it is ignored, and in the documentation of this practice, only the isolated threads of decentralization's positive impact get covered, ignoring many of its larger, more important, implications.

"Community Participation." Participatory planning, community action planning, community-based development, community needs assessment, and so on, have become an integral part of today's planning jargon. Unfortunately, however, these terms are often just smoke screens for "nonparticipation." We use a few examples to illustrate this.

In Swaziland, by law, every project must have an environmental impact review (EIR), and every EIR must have a community participation component. This may sound credible on paper, but let us take a case in practice, an ambitious project (about $800 million) to build a bypass road around Mbabane. This was a massive undertaking, involving highway construction around the inner city which resulted in large-scale displacement of local communities. "Participation" under this project included the usual feasibility study during the design phase, essentially a physical survey to determine the need for compensation and resettlement. This was followed by a community meeting in each affected settlement prior to implementation to "inform" the people of the project impact, allowing little room for community input or feedback. As reported by members of one of the affected communities, "We are not allowed to discuss or protest."

A pilot upgrading project in Swaziland, implemented in 1999–2002, suffered from similar problems. A strong and very extensive community participation component was built into the project design, with project outreach facilitators selected from the various communities to build the bridge between the local residents and the implementing agency. An agreement was reached in the initial phase of the project, with regard to the level of services to be provided, and the related costs to be borne by the community. However, due to a systemic delay in the project cycle, and subsequent price inflation, the agreed-upon plan was changed, without consulting the community. The people, as a result, felt angry and cheated, and refused to pay taxes or payments two years after project completion. The government, on the other hand, complained about its inability to recover costs, and the "unethical tactics" of the people to avoid paying.[10] The lesson here is that participation is not really participation unless it goes all the way through. Here, the project was well-intentioned from that aspect, but one slight slip along the way was significant enough to undo all the efforts.

In China, the approach of local governments toward participation, and development in general, may be summed up by the common belief that "the people must sacrifice for the larger good." Unfortunately, however, the ones "sacrificing" are typically the poorer sections of society who have little room to negotiate or voice their opinions, particularly on the issue of land acquisition and compensation. Further, while participation may be outwardly encouraged, the constraints to free speech are very real. An example of this in Mathema's experience was the interference of local officials in community meetings, who often "planted" unrecognized members of their staff in group discussions aimed only at community members to note what people were saying. This was despite requests for direct face time with the communities without interference from the officials who may directly or indirectly intimidate the locals from voicing their opinions. This is what we mean when we say community participation is often only on paper and not in practice.

PROJECT PROCEDURE: THE LOGICAL DISASTER

At the root of many problems being experienced in development today, and especially in urban development and housing, is the system which has emerged over the last forty years for project identification and development. There is a logical thread linking all stages of the development process which starts the moment a project is identified and which inhibits reflection, consultation, participation, or even its cancellation. Careers of project management staff (whether from international development agencies or governments) are built on the twin foundations of speed and efficiency: to these people grassroots involvement is seen as both a threat and an irrelevance. At best, it is seen as risky. Thus they are much more comfortable developing a project along the conventional (and they would say logical) process outlined.

Projects must be based on certain assumptions about the target beneficiaries, the amount of money to be spent, and the ultimate objective in terms of services to be provided: at no stage in the subsequent process is it easy for those basic assumptions to be challenged. Indeed, the duty of the designers of each phase is to *build on the work of the preceding stage*.

At the risk of repeating the obvious for technical people, it is useful to outline the process typically followed:

The project sponsor will start by preparing terms of reference for the proposed project. These will typically include budget ceilings, intended scope of work and standards to be developed within the target community, implementation time frames, and so on, but are prepared *before* any surveys have been prepared or interactions with the community have taken place. Thereafter, the sequence of events runs like this:

Appoint consultants to undertake feasibility study. Consultants will be required to demonstrate an ability to respond to the terms of reference for the study, and their experience in similar projects.

Surveys undertaken. The successful consultants then embark on a feasibility study including a socioeconomic survey. Questionnaires will be administered regarding the residents' expectations, demands, willingness to pay, and so on.

Outline proposals prepared and submitted for project appraisal and budgetary approval. These will amplify the solution outlined in the original terms of reference, giving cost data and physical form to the proposal. Insofar as the responses from the questionnaires are used, they will be used to validate the solution proposed. Have projects been sent back to the drawing board at this stage? The answer to this question speaks for itself, because all parties have, by now, committed themselves to speedy design and implementation.

Appoint consultants (2). At this stage, using the outline proposals as the basis, a new set of consultants may be brought in: typically engineers who will take responsibility for the design of the infrastructure. The parameters within which they must design have already been established in the feasibility study.

Prepare final cost estimates and obtain funding approval. With detailed engineering complete and final costs obtained, the project can be packaged for final funding.

Prepare tender documentation and go to tender. However, funding availability often cannot be taken for granted, and there may be a considerable gap between the time when designs are complete and funding is approved. But whether the next stage is one month, one year, or five years later, the next stage will be to prepare the tender documentation.

Appoint contractor. Following the receipt of tenders, a number of decisions must be taken. They can be particularly difficult where the costs are far higher than the estimates. Will it be necessary to obtain additional funding before proceeding? Can the standards or level of provision be reduced? Should the work be done in stages? These decisions must be taken quickly and often a revised scope of work is negotiated with the successful tenderer (not the beneficiaries).

Supervise construction and approve completion (Shock 1). For the sake of argument, we can assume that these procedures are followed efficiently and that the work is under construction. The residents are usually the last ones to know exactly what will be done and where.

Hand over (Shock 2). This is now the stage where the community may be in for another shock. What has been provided may be quite different from what they expected. For example, they will recall the survey in which they were asked what improvements they wanted, and they might have said tarred roads and street lights, but the contractor leaves and the roads have not been tarred. Or they asked for individual taps, but only standpipes have been provided. Why?

Payment (cost recovery) (Shock 3). It is not long before public meetings are held. Mayors and councilors address the people to inform them that they will have to start paying for the improvements, and each household will get a monthly bill.

In brief, the system of project design and development inhibits and sometimes prevents flexibility and responsiveness. By creating deterministic linkages between each stage of project preparation, there is an inevitability about the result. If the initial assumptions are wrong, then the result will also be wrong. But most importantly, the course on which the project ship has sailed cannot be easily changed by the users. Indeed, such inputs create uncomfortable questions and expose the project team to allegations of inefficiency, delay, "not knowing what they are doing," and so on. Unless and until development is conceived in a fundamentally different way, users will have little say in what is provided for them, and how.

REALITY CHECK

It should be self-evident that those who work on development projects should have at least a basic understanding of the situation on the ground through direct contact, and not through just secondary sources of information. Unfortunately, we can cite examples of far too many people who work in the housing and infrastructure sector who have never ventured into what they call "slums" or those who make only token visits to these areas in four-wheel drives, like a trip to the safari park. Only very few actually spend time with the people and try to understand their priorities or needs firsthand.

Just like a doctor who cannot diagnose or treat an illness effectively without touching the patient or monitoring the recovery, how can one know the real situation in these communities without at least talking to them? It is this absolute necessity for personal contact, and on-the-ground knowledge, that we stress here as a critical component of development practice— right from the assessment phase through to project implementation and follow-up monitoring and evaluation. We illustrate this here with some personal experiences, which also show that gauging the actual problem is not actually that difficult, or costly, or time-consuming: it just requires talking directly to the people involved, just like in the medical or any other field. Surprisingly, though, it is rarely done.

The first example comes from Kenya. Most people in the development sector in Kenya will tell you that there is no dearth of NGOs working in the social development sector, including job training, and health. It may be true that many NGOs are active in Kenya, but one need just visit the informal settlements to see that many gaps still exist. Here are two cases which Mathema stumbled across while doing interviews of residents in the informal settlements of Nairobi in 2005.

- The first is a 50-year-old who lives alone in a small rental shack in an informal settlement, with no job or income other than nominal support from a brother to cover her rent and basic food. One day, she gets very sick with diarrhea. There is a newly built (donor-funded) medical facility in the area, we are told, which was recently inaugurated with much fanfare, and provides 'cheap'—often 'free'—medical treatment to the poor households in the community. When the patient in question is taken there, the nurse prescribes three laboratory tests and a 15-day dose of antibiotics, costing some KSh800 ($10), equivalent to three months of her income. This, for a simple case of diarrhea. This shows that while the facility exists, it is in fact unaffordable to the vast majority of the poor, as a result of which they are often forced to resort to home remedies, self-treatment, or no treatment at all. So who does the facility cater to in reality?

- The second is a 38-year-old recovering alcoholic. He was unemployed at the time of the interview, and expressed interest in kicking his addiction and enrolling in a job-training program. There is a "well-known and reputable" NGO in the vicinity running training programs for youth which, we are told, provides just the sort of service sought by this individual. A call is made to the NGO to inquire into the prospects, and as it turns out, the NGO provides vocational training to youth within the ages of 14 and 21 years, thus excluding him and all others like him from any formal opportunity for reform. This, when in fact, drug and alcohol abuse among the youth in Nairobi is highly problematic, and most prevalent in young unemployed graduates in their late twenties and thirties. So, here too, what appeared to be an inclusive training facility ended up being exclusionary in practice.

The point here is that the types of instruments commonly used to assess needs of low-income communities typically don't involve the one-to-one communication with the people that is necessary to really understand the situation on the ground. For example, quantitative surveys can never reveal this type of qualitative information, and hence the real story often gets overlooked. To really understand people's needs requires talking to them directly.

Another example comes from Swaziland. In conducting a country-wide infrastructure needs assessment study of the peri-urban areas of Swaziland,[11] we were told to focus on the physical aspects of these settlements, under the premise that the broader social issues such as AIDS were being adequately addressed by NGOs. Still, as social and economic conditions are direct determinants of access to physical services, we included a question about the social

Box 4.9 Reality in the field.

Graves in the informal settlements of Swaziland. We were told, "People here are dying like flies; there is no place left to bury our dead!"

Source: Field research by Ashna Mathema in 2002–2003.

problems prevalent in these communities. In talking to the people from each of the 52 settlements visited, it became clear that NGOs had been largely ineffective in addressing the HIV/AIDS issue, both in terms of awareness creation and encouraging precautionary measures, which the government had so far been unaware of. The result of that negligence is evident today: Swaziland has the highest prevalence of AIDS—as a percentage of the population—in the world (see Box 4.9).

SUPERFICIALITY OF INDICATORS

Quantitative indicators are often used as substitutes for qualitative benefits, rather than a means to quantify the qualitative benefits. Numbers are a good way to quantify success, but this brings us back to the fundamentals: what is success? And success for whom? A project is often deemed successful if certain indicators are reached; what happens outside of those indicators often has no bearing with the success rating of a project. In reality, however, these indicators are mere numbers, and numbers can be manipulated to tell any story. The only way to give meaning to these numbers, that is, to "contextualize" them, is through closer interactions with the communities.

Here we cite another example from Nairobi, a site-and-services project initiated in 1976 that was successful according to the indicators but deemed a failure in the eyes of many of the beneficiaries 25 years after project completion. But does anyone care, and does it matter? We contend that it does, and that something needs to change in the way success is measured. But first, here is the story, and it comes from just three days in the field interviewing the beneficiaries of the project to get a sense of how the project fared (that is to say, that this sort of fact-finding on the ground is not tedious or time-consuming). These are some of the prominent common themes which don't feature in any of the project evaluation documents.

People reported that corruption was rampant in awarding construction loans for housing: loan recipients were made to sign for amounts as much as KSh5,000 ($710) more than they actually got from the loan officer. This was 20 percent more than the average loan size of KSh25,000 (approximately $3,000) and the amount was "retained"—in cash—by the loan officer as "fee" for "getting approved." As one interviewee reported,

> You should have seen the stack of money under his chair! Five thousand shillings times the number of loan applicants is a lot of money! And most of us wanted a loan . . . Further, despite the fact that there were sufficient funds to give these small construction loans to all the plot allottees, not everyone who wanted a loan got it.

That was during project implementation. Since then, much more has happened. Residents allege that, one, the City Council, the implementing agency, has been involved in the illegal sale of land reserved for public uses in the settlement. As a result, there are encroachments on playgrounds and roads, and people have built on sewer lines and drains, which has led to backflow and flooding problems in certain areas.

Two, according to the terms of the agreement, the beneficiaries were to be awarded titles or leases after repayment of the loan (over a 25-year period). However, this did not happen. People who had paid up their loans were told to "come back with more money." First it was for various "processing charges" (legal fees, surveyor fee, etc.). Now, people claim that the City Council has imposed an additional requirement of KSh50,000 ($7,140) per plot before the titles can be issued. No one knows the reason for this extra cost, but they cannot dare to question it. In other words, after 25 years of compliance with the terms of the loan contract, the beneficiaries still have no security of tenure, or the ability to trade and profit from their hard-earned and well-deserved property in the open market.

So, how is it that none of the project documents make mention of any of this? It could be one of two things: either the monitoring and evaluation process was inadequate, because if it were adequate, these issues would have surfaced without too much effort. Or that it was adequate but the correct information was suppressed. And this has to do with how the "performance" of the project, and consequently that of the project manager, is judged in general. To show faults in the project—which can be critical determinants of ensuring success for follow-up projects—is almost never in the interest of the project manager, because it reflects poorly on him or her. This is another systemic fault in the incentive structure of donor agencies, but we will not get into that here. Let it suffice here to say that the project was closed, and in the books of the donor, it was a success based on the "indicators" that quantified how many people got houses or plots or water connections, and so on. And while we agree that the people are probably better off than they were 25 years ago, the fact remains that they got a raw deal, and there is no one who will stand up for them after they have gracefully fulfilled their side of the commitment.

FROM POLICY TO PRACTICE: THE MISSING LINK

Based on these observations, we conclude this chapter with what we believe are some of the primary reasons for the missing link between planning and practice.

First, to find the right solution, one must first be able to define the problem. This requires planners to carry out assessments and work with the residents to propose feasible solutions.

Second, every decade has seen one or the other sector take the center stage of the development agenda. The 1990s saw a shift towards governance-related issues: anticorruption, democracy, and good governance, and a waning of interest from the housing sector. But a fact remains: poverty is directly correlated to access to basic services and housing, which in turn is a determinant of human development. In other words, housing is a basic need, without which no amount of anticorruption or democracy building can be sustainable. So housing does deserve much more attention that it is currently getting. Another point in this regard is the approach to housing. Despite the rhetoric about government being a "facilitator rather than the provider" of housing, governments across the world still continue to build public housing based on the now-rejected Western models of the 1970s.

Finally, donor money may be well-intentioned, but it is often channeled through national governments to develop large "visible" megascale projects that lack attention to detail and are often fraught with corruption. Community-level efforts are more difficult and seem potentially expensive to administer, but history has shown that they prove more worthwhile in terms of the net impact per unit of expenditure. People know best what they want and how they want to get it. So, much more emphasis needs to be placed on building consensus with the beneficiary communities.

And that leads us to the final point, which is our emphasis on the fact that no one way is the best way, and that what is a "best practice" in one place could in fact be viewed as completely preposterous in another. *Process* is key, from the start of the project right to the very end, from the assessment to implementation and follow-up. Easier said than done, but if we can get the process to be more responsive to the people and their needs, the rest can easily be made to fall in place.

In other words, what is needed to close the gap between planning and practice is greater emphasis on the assessment phase of projects, based on real field assessments involving direct interaction with the people, a stronger link among the physical, social, and economic components of development, and recognition of housing as a basic building block of human development. Once this is done, people will find the solutions largely on their own, reducing the need for large, unmanageable and corruption-ridden capital investments, thereby making the interventions cheaper, more effective, and far-reaching.

5 The Legal Framework
Oppression or Defiance?

This chapter is about the ladder to legality faced by the homeless, and the different types of control which face the homeless in their search for shelter. It documents the array of legislation which impedes them from climbing the ladder to acceptance. It also details the type of legislation and regulation affecting them, and looks at both the cost of complying and the cost of not complying in terms of criminal sanctions. The effect of this legislation is that those with homes keep out those without, by means of rules which the homeless are legally required to follow but cannot.

It examines the need for norms and standards, their scientific validity, and the role that they should play in the control of development. It shows how, unless they are appropriate for the economic and cultural circumstances of the community, they become weapons for control instead of means to improve public welfare.

The law is a tool used by human societies to protect themselves from antisocial behavior. In stable societies not affected by rapid change, knowledge of the law is embedded in both values and practice. Those who transgress therefore know that they are doing so. In rapidly evolving societies, and ones in which people from different backgrounds coexist, the situation may be very different, and people find themselves criminals without any understanding of why this is so.

Similarly, in environments in which there are comparatively few actors—the typical rural community—life does not require regulation. The opposite is true in the highly complex environment of the big city. We can draw a distinction between, on the one hand, the common and virtually universally accepted code of ethics which exists in all religions and is, so to speak, taught from the mother's knee, and, on the other hand, the petty regulation which is premised on administrative necessities and/or the social values of the lawmakers.

An example of the former is the prohibition against stealing; an example of the latter is a regulation regarding keeping off the grass in a public park. In terms of the homeless, the former would include occupying other people's property without their consent, and the latter would be sleeping overnight on a railway station bench. In the continuum between these two extremes there lies a huge range of moral and practical alternatives.

Before going into detail, we should step back and understand the nature of standards and the procedures by which they are formulated and applied.

On the one hand, there is a strong commonsense component to standards and standard setting. All professions have certain standards, and building is no exception. A look back at the horrors of uncontrolled development during the industrial revolution in Europe and the urbanization in the United States that we described in Chapter 1 shows the need for some controls by society to protect its citizens from danger and disease and to promote their welfare.

So it would seem that there must be an element of objectivity in standard setting which, although there may be some regional variations, embodies a common rationale. For example, to achieve a given level of natural light defined in terms of the international standard of lumens within a room requires larger windows in the gloomy skies of northern Europe than it does, for example, in brilliant light of the Sahara desert. Standards concerning the strength of walls should also be objectively verifiable.

However, when we look at actual standards we have to wonder. For example, in the field of minimum plot size there are such huge variations that it is hard to imagine what objective standard they could be based on. What is even more startling is that the standards within specific countries are held up as absolutes (see Box 5.1).

Similarly strong positions are taken about minimum house size, maximum permissible density, building heights, plot setbacks, rights-of-way, and so on. Politicians make a strong point of insisting that a minimum house size is necessary to prevent overcrowding and to provide people with decent accommodation. For example, in the early days of independence of Zimbabwe, the minister then responsible for housing insisted that nothing less than four rooms should be developed.[1] This was unaffordable to those below the median income and therefore could not be financed by the USAID's housing guaranty program. It was three years before the penny dropped and this insistence on space standards was dropped. Thereafter, a successful sites-and-services projects was implemented at scale. Participants built small houses to start with, but over time the majority have now blossomed into little mansions.

Similarly, caps on density are put in place to ensure that the supporting infrastructure is adequate. This was done in Mumbai, India, in the 1960s and 1970s, where the government drew up a master plan with all the right regulations (floor-area ratios, etc.), catering to a maximum population of seven million.[2] This did not stop the city from growing, however, given the business hub that it is. Today, the city has a 20-million-strong population, but the regulations remain the same as they were in the 1970s, as does the infrastructure. As a result, 55 percent of the city lives in slums, and the infrastructure, even in the formal areas, is seriously overstrained. So what

Box 5.1   Relativity of absolutes.

One of the interesting features of the politics of development is the passionate positions taken in many countries regarding plot sizes. The experience of Zambia in this respect is quite funny. When I was involved in negotiating a loan from the World Bank for the Lusaka Squatter Upgrading and Sites and Services Project, the Bank took an early position that the standard sizes (12m x 24m = 324m^2) in Zambia were excessive. They pointed to the impact on infrastructure costs, and ultimately the cost of such low densities in terms of journeys to work etc. The Government was adamant that this would be a point on which they would not budge in spite of very reliable data that the plots were not being well used.[1] In the end the Bank conceded the point in the interest of progress.

However, when I came to implement the project, I found that the residents of the settlements to be upgraded had a different view. Their average plot size was about 150m^2, and when they were offered plots of 210m^2—which had been shown to them on the ground before making any infrastructure developments—they were delighted. When, subsequently, the Minister took a tour around the settlement and saw the plots (not knowing what size they were) he too was extremely satisfied.

Thus it is that standards based on fixed numerical values can play tricks on us; and for anyone practicing in Africa it is impossible to even contemplate the 25m^2 plots being used in India, or for that matter, the 50 m^2 plots used very successfully in Manila. Yet, in a different project in which I was involved[2], a fire in the settlement prompted the mobilization of emergency aid to the occupants of squatter shacks that had been razed by fire, when we provided a simple solution of firewalls spaced 6 metres apart, on a plot of 60m^2, against which (or between which) people could build their houses, the beneficiaries were delighted and made no demands for the prevailing standard minimum plot in South Africa of 330m^2. In that, they had a point: in Holland a 6 meter wide plot is not unusual, and five storey buildings built on such plots, housing three dwelling units, were the standard in the early 20th century. Today these houses are admired and sought after.

But in many developing countries, especially in Africa, such standards are illegal. For example in Accra, Ghana, the minimum plot size is 450 m^2, minimum plot width is 15 meters, and maximum plot coverage 50 per cent.[3] Similarly, in Lagos, the official minimum plot size is about 1000 m^2, and 650m^2 in so-called "high-density" areas.[4] According to a senior government official, this is based on the fact that "Nigerians like a lot of land," but an unintended consequence of this regulation is that nearly three-fourths of the city has no access to land, as a result of which slums have proliferated along the coastline and other hazard areas. (RM)

[1] R J Martin, *Gardens and Outdoor Living: Research Study No. 1* (Lusaka: National Housing Authority, 1974), which demonstrated that only about 42percent of the households used the plots for any gardening activity, and that only 1.5percent of the surveyed households used more that 65 percent of the 324 m^2 plots. This data did not weaken the government's resolve to stick with its minimum plot size.

[2] Under USAID's Community and Urban Services Support Project; the settlement concerned was Duncan Village in East London, and the design and implementation of the emergency project was undertaken by Anthony Fauld Mann, a consultant to the project.

[3] Government of Ghana, L.I. 1630: National Building Standards, 1996, section 13.

[4] Mathema, A. (2007). "Slums and Sprawl: The unintended consequences of good regulation," The World Bank.

did those regulations really achieve? Just government apathy in addressing the growing infrastructure needs, masked by helplessness in the face of an uncontrollably growing population, particularly the "illegal" informal slum dwellers, choking city services.

That said, even if the actual regulations are imperfect in the way they interpret and apply a scientific standard, is there a real need for such regulation? The general consensus is yes. It is interesting that in a survey undertaken in Zimbabwe, 89 percent of the respondents said that standards are useful. We must presume that the respondents represented that population as a whole, so it is not surprising that the most cited reason for approving of standards was that they "enhance the aesthetic value of housing as a consumer good."[3] Control was also considered important by 28 percent of the respondents. On the other hand, public health was only considered important by 9 percent.

However, if we were to segment the population into those who could afford the standards and those who could not, one must wonder whether the results would be the same. The general consensus is that more than half the population cannot afford a dwelling that conforms to the official minimum standard. The figures just cited tend to suggest that it was the response of the "haves" to regulate the behavior of the "have-nots." This is a far cry from the days when public health was the primary motivator in the application of building standards in the face of exploitation of the poor by landlords. This passage puts the situation well:

> In (developed countries), especially from the 19th century onwards, standards came to be required as a means of protecting the weaker members of the community. In particular, they protected workers from the rapacity of landlords who were tempted to produce rapidly some form of shelter that did not meet the needs for hygienic living, safety or privacy, for which they charged exorbitant rents.[4]

The same passage contrasts this with the situation in developing countries where the motivations were, and seem to continue to be, different:

> In developing countries, standards for shelter provision seem to have arisen as a means of protecting the wealthier and more educated minority in a country against the incursion of lower levels of shelter provision that are within the reach of the minority of the population.[5]

Thus we are repeatedly confronted by the use of standards as an exclusionary tool, either by adopting inappropriate and even arbitrary standards, or inappropriate ones. The net result is that a significant proportion of the population does not, and indeed cannot, comply. Where they comply, the objectives underlying regulations are flouted, as is evident, for example, from the many slums in Nairobi, Accra, and Addis Ababa, in so-called legal subdivisions. Here, planning guidelines are followed up until the plot level, but flouted completely *within* the plot: far too many families share the plot, resulting in extremely high densities, overcrowding, and highly inadequate basic services.

In this context there are several pertinent questions to be asked about standards and regulations.

1. Are they a product of the community in which they are applied?
2. Are they affordable by those to whom they are intended to apply?

Any review of building standards in developing countries demonstrates vividly the fact that the answer to these questions is in the negative.

There is a third question to be answered:

3. What benefit would be gained by the occupants if the regulations were applied?

If we look back, once more, to the late 19th century, it is clear that the answers to the preceding questions would have yielded a positive response to questions 2 and 3. It is undeniable that, in spite of the howls of protest from the factory owners and landlords of that period, housing standards increased dramatically as a result of the interventions of the state in the form of building and public health regulations.

The situation is simply different today. The gaps between affordability for housing and the cost of so-called minimum standard housing are so large that the 19th-century response is not replicable. Many governments have refused to believe this, and have bravely proceeded to develop housing projects meeting these minimum criteria, only to find that people cannot afford to occupy them without subsidies, and that government itself cannot afford those subsidies. The case of Addis Ababa, which is illustrated in Box 4.1, shows how standards in redevelopment projects act as exclusionary forces. In the new projects being built by the local authorities, not even a one-bedroom unit is affordable to the bottom 65 income percentile.

Equally important to note is that the health angle, which took predominance in structuring the early 19th-century housing regulations, seems to have taken a backseat. Rather than using regulations as a means to an end—that is, better housing to achieve better health outcomes—housing has become an end in itself. Standards are often set to achieve better aesthetics and uniformity, rather than to remove or reduce externalities associated with poor quality housing and infrastructure.

Let us move here from the perspective of why standards are needed to what the man in the street may want and can do. If we start from the perspective of a person, whom we shall call Ali, who participates in a land invasion and builds himself a house, the sequence is typically to start by building a single room.[6]

As his position stabilizes he may gradually save enough to build a second room, which he then rents to another family. With the income from this rent it does not take long before he can build a third room. Meanwhile, he has established a business—let's say he is a carpenter— and starts making furniture in the backyard. He may also, for example, use one room of the house for his wife to ply her trade as a hairdresser. In the examples that we know best, the technology adopted in the construction of these rooms will initially be of the most basic nature. Secondhand materials, or materials found on site (e.g., mud), will be used to make an enclosure of an absolutely minimal nature. It will typically have very small windows (large windows being expensive, and making the unit vulnerable to theft) and minimal insulation. It will probably leak here and there. But it provides cheap shelter, thereby saving Ali the cost of renting, and (in our scenario) allowing him to obtain a modest income from renting out rooms. Moreover, by saving on expenditure on housing, the family is able to save to develop a business.

This account will sound familiar to many readers, as will the pictures of shacks and slums which are used for their shock value in international development literature. But do the occupants of these settlements see matters in this light?

The case studies which we used in the three preceding chapters make one very important point: we are witnessing a rapidly evolving environment. Thus to outsiders an informal settlement may look like a city of shacks; to most of the occupants it is a city of achievement and change. From the first moment when they occupied empty land to the present time when they see the homes of hundreds, or even thousands, of families, they see progress. Every year there is more progress, as houses are enlarged and improved. The simple mud structure is replaced by concrete blocks; the one-room house is now a four-room house. Some families now have their own water connection. Others have planted trees, or have vegetable gardens. In this way they see their present, imperfect environment as a step on the ladder of self-improvement in terms of housing as well as economic opportunities. Would the imposition of standards to such development have helped or hindered this process? Certainly, if you look at the conventional bundle of regulations and standards it is not difficult to conclude in the negative.

We should ask whether Ali has done anything wrong by housing himself and generating employment. From the point of view of the national economy, surely he is contributing in a very positive way. He has used his own capital to make fixed investments, he has relieved the state of the obligation (which many impose on themselves) to provide housing, and he has contributed meaningfully to the creation of employment, both in the act of developing the settlement and in the economic worth of the informal sector in which he may work.

The following example illustrates the efficiency and effectiveness of community-based development which follows cultural as opposed to statutory standards.

Research in Zambia[7] recorded not only the design of houses in detail, but also the changes which the occupants had made over time. In having the freedom to make these changes they were able to respond to their needs in a cost-effective and prompt way. Box 5.2 shows the effectiveness of the residents in meeting their needs, as well as the inappropriateness of the regulations: none of the housing illustrated here would be allowed if the letter of the law were to be followed.

There are three aspects which are worth noting.

Efficiency: The plans reproduced here show how close a degree of fit was achieved. Note, for example, how the width of bedrooms is based upon the length of a bed. Other aspects which are not immediately clear from the plans are:

- The rooms are 30 percent smaller than the officially permitted minimum.[8]
- The interior of the house is used primarily as security space during the day (i.e., for keeping things in) and for sleeping in during the night. Daytime living is largely out of doors: even sleeping—the afternoon nap—takes place outside, in full view of the neighbors.
- Planning takes place in three dimensions. In other words, if one was to draw a plan at different levels in the house one would find different uses. The ground is used as a depository for pots and pans, bed-mats, chairs, and tables. The upper level of the house is used as a wardrobe with clothes hung on a string across the room, like a washing line,

Box 5.2   House plans for Mwaziona, Lusaka.

1 In these plans note how closely the room size is tailored to the length of a bed.

4 An example of one of the houses in the group studied (Case 1).

5 Case 2.

and pots hung up. The roof is used as a third storage area for large objects such as maize, bicycles, and so on.[9]

Flexibility: People living in informal settlements experience a more rapid rate of change in terms of economic circumstances and family size than does the population as a whole. Their housing illustrates this more clearly than any other aspect. In Box 5.2, the two lower rows show the evolution of two houses over a period of seven years.

Case 1. The house was built in 1968 at the cost of one month's salary and occupied by two adults and five children. By 1972, the house had been sold to a man and wife with two children and had been changed. By 1975, the house had been completely replaced by a stronger (partly concrete block) house at about four times the cost of the first one. The family is now two adults and four children.

Case 2. The house was built in 1968, when it was occupied by only one man. By 1972 it had also been sold, in this case to a family of two adults and two children. This family added on two rooms. By 1975, modifications had been made to the plan: blocking off doorways, opening up new ones, and so on, making it suitable for occupation by an additional small family (brother and his wife).

The rate of change in these cases is due largely to the ease with which changes can be made: this is due, in turn, to the use of a walling material which is, effectively, free: sun-dried mud blocks.

The plans of the area (Box 5.3) show the changes of the whole environment over this period: latrines and bathroom structures come and go; houses move; they change. It is rather like watching the evolution of the species over a very condensed time frame.

Box 5.3   Spatial evolution over seven years.

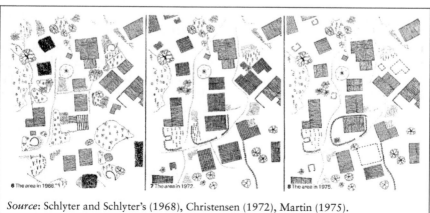

Source: Schlyter and Schlyter's (1968), Christensen (1972), Martin (1975).

Economy: The people's resources are very limited, but they make significantly more space for their money than in conventional solutions, and the houses themselves are relatively better. The ten dwellings surveyed and illustrated here cost the same as a single so-called low-cost house built to the standard room sizes laid down by the Public Health Act, and to the construction standards laid down by Zambia's National Housing Authority for Very Low Cost Housing.[10] Ten houses for the price of one. This is a very remarkable fact. How is it done?

Wherever possible, free materials are used. The most important of these is the soil.

- As in the planning of the house, so in the use of materials: nothing is overdone. For example, instead of the 100-mm floor slab that we often use in residential buildings, a screed of 15 mm is used. If laid on a strong base, and if the furniture is light and the room is small, this is adequate.
- The expensive items such as doors and windows receive special attention. Doors are rarely more than 650 × 1700 mm, 25 percent smaller than the standard, while glass windows are never used. Instead, ventilation openings with wooden shutters are used, with an area, on average, of about 4 percent of the floor area.[11]
- All structures are kept at a noncritical level. For example, if roof spans are small, secondhand timber from packing cases, and so on, can be used.

The most important commodity is space: to make the enclosure out of permanent materials is of secondary importance. If the initial enclosure of space has a useful life of three or four years, and is paid for (as is normal) during construction, those three or four years gives the householder time to save for something bigger or stronger or both.

There is another aspect of economy: land use. In the example illustrated, although there is considerable space between the buildings, the density is high. There is about 90 m² of private space per dwelling, at least three times the built-up area, but yet the net density is about 37 houses per hectare, more than double the density of housing layouts approved by the Zambian Ministry of Local Government and Housing. This is achieved by the very modest 10 percent of the land surface going to circulation, which can be achieved by an overlap between the circulatory and private open-space functions.

THE LAW

Examining the small print, so to speak, of the law reveals a situation which is very different. Those who do not comply with regulations are considered guilty of a criminal offence. Thus far from being lauded as productive members of society, the poor are typically vulnerable to a multitude of controls and sanctions, including fines and the demolition of their house. It was on exactly these grounds that 60–70 percent of the population of Mbabane, Swaziland's capital, was deemed "illegal." "They do not follow the law, they build as they please without following the regulation, and hence have no rights . . . they are illegal," as a local official put it. And this is exactly the point we are trying to make: How appropriate is a law which relegates the vast majority of the population to the other side, resulting in mass noncompliance? Is there not something inherently wrong with a law established on grounds of protecting the rights of citizens that, instead, demonizes the way of life of a vast majority of its residents?

The term *reductio ad absurdum* is used in logic, but it has a particularly apt ring when we look at the potential penalties which may be applied to the person we have just referred to, Ali, who builds himself a four-room house, one of which he occupies, two of which he rents out, and the fourth of which is used as a hair salon. The table in Box 5.4 shows the penalties to which he could be subject if he were to try it in Johannesburg.

We should note that most of these fines can be imposed for each day during which the owner of the building does not comply with the relevant bylaw.

This example is intended to demonstrate that the law is, as cliché would have it, an ass. But if you are a resident of an informal settlement, laughing at the law is of little help when municipal policemen threaten you with such sanctions (see Box 5.5). And it gets worse: often, in their zeal to enforce laws, the authorities resort to violence. The destruction of houses and property and often the injuring, maiming, and death of some residents are seen as necessary adjunct of the proper enforcement of the rules of society.[12]

There is another, and serious, implication. The fact that residents do not comply with the regulations and bylaws gives those in authority a tool with which to extract bribes. In practice this may be the more serious threat to a householder's plans to build or improve his house. It provides the politicians, police, and civil servants with a perennial opportunity to threaten so-called offenders, and the opportunity to intimidate residents with threats of fines and/or demolition. Thus these outwardly harmless regulations become tools for oppression and victimization.

With the demolition of your house goes the loss of your savings as well as shelter. The next logical step is, of course, for people thus threatened to try to bribe their way out of the fix that they are in, and thus begins the slippery slope of mal-administration, exploitation, and corruption by those in authority over those whom they are supposed to protect.

Box 5.4   The cost of breaking the law.

Legislation	Offence	Punishment* Fine (Rands)	Prison Months
Town Planning	Failure to comply with provisions of an approved town planning scheme	Court order to demolish	Not stated
	Failure to remove illegal structure	Cost of demolition by council and all associated legal expenses	Not stated
	Contempt for tribunal	R 2,000	6
Building regulations	Foundations not complying	R 4,000	12
	Walls not complying	R 4,000	12
	Room size not complying	R 4,000	12
	Windows not complying	R 4,000	12
	Sanitation not complying	R 4,000	12
National Health Act	Creating pollution detrimental to health	R 10,000	60
	Constituting a health nuisance	R 10,000	60
City by-laws	Failure to connect to municipal sewage system	R 500	-
	Failure to keep backyard and/or refuse area in a sanitary condition	R 500	-
	Installing septic tank without a permit	R 500	-
	Pollution of a catchment area in a way that creates a public health nuisance	R 1,500	-
	Failure to provide an adequate potable water supply	R 500	-
	Use of water obtained from a source other than the municipal water supply	R 500	-
	Operating a salon without permit	R 500	-
	Operating a salon on premises which do not comply	R 1,000	-
	Salon waste not disposed of in approved manner	R 500	-
	Operating an accommodation establishment without a permit	R 1,000	-
	Failure to provide towel rails in bathroom	R 500	-
	Incorrect storage of dirty bed linen	R 500	-
	Incorrect storage of clean bed linen	R 500	-
	Failure to keep sanitary ablution and water supply fittings in good working order	R 1,000	-
Water bylaws	Unauthorised connection	R 1,000	6
Electricity bylaws	Unauthorised connection	R 1,000	6
	Total	R 53,500 ($8,250)	198

Sources: City of Johannesburg Metropolitan Municipality: Public Health By-laws, Provincial Gazette Extraordinary, No 179, 21 May 2004, Johannesburg; Government of the Republic of South Africa: National Building Regulations and Building Standards Act 103 of 1997 and National Health Act, 2004.

Box 5.5 Demolition police in an illegal settlement in Addis.

Demolition police on the watch for "illegal" construction.

Source: Fieldwork by Ashna Mathema in 2004.

PROCEDURES AS BLOCKS TO DEVELOPMENT

One of Hernando de Soto's most important contributions has been to demonstrate what a negative impact bureaucracy can have on development.[13] The multitude of petty regulations, all of which may have seemed essential at the time that they were developed, and which (for the sake of argument, shall we say) may be applied in good faith, amount to an obstacle course for the poor. This regulatory minefield has one of two effects: taking the example of house construction, either people simply ignore the regulations and develop anyway; or the regulatory process adds a significant cost which reduces the amount available for construction.

Home builders in South Africa (which in many senses prides itself on pro-poor policies) face similar problems. Even in the well-managed upper income areas of the city of Johannesburg, granting of the right to subdivide a portion of land in an area (where subdivision is officially encouraged under a policy of densification of older suburbs) can take over two years.[14] These procedural delays constitute a clear incentive for developers and others to ignore the law. How much more difficult is it for low-income people to comply? In the case of low-income housing, the procedure shown in Box 5.6, published as part of a manual on how to implement low-cost housing projects, speaks for itself.

The complexity and the degree of professional expertise necessary to undertake these tasks mean that it is clearly impossible for a

Box 5.6 Procedure to be used in development.

1. Securing rights to the land
   - Identify land
   - Establish registered owner
   - Negotiate with owner
   - Conclude agreement
2. Land investigation into legal/cadastral position of land
   - Title deed description and area
   - Existing leases (registered and unregistered)
   - Restrictive conditions/servitudes
   - Restrictive conditions—other rights
   - Restrictions—surrounding development
   - Environmental impact study
3. Town planning layouts etc
   - Obtain base mapping
   - Define planning parameters
   - Prepare draft layout plans
   - Test plans against engineering requirements
4. Township establishment
   - Prepare re-zoning/subdivision application
   - Submit application to local authority
   - Advertising of application
   - Local authority approval process
5. Land surveying
   - Collate base information
   - Outside figure survey
   - Preliminary calculations
   - Fieldwork
   - Prepare draft General Plan (GP)
   - Submit GP to Surveyor General
   - Initial examination of GP
   - Advise amendments/corrections
   - Amend/correct GP and re-submit
   - Final examination of GP
   - Approval advice for GP
6. Preparation of Conditions of Establishment
   - Draft conditions of establishment
   - Approval of conditions of establishment
7. Compliance with Conditions of Establishment
   - Preparation of application
   - Advertising of removal of restrictions/court order
   - Advising deeds office
8. Opening of Township Register
   - Drafting of application
   - Submission of approved GP to Deeds Office
   - Lodge documents at Deeds Office
   - Conveyance procedure
   - Registerable plot

Source: Department of Housing/National Business Initiative: *Housing Project Programming Guide* (Pretoria: Dept. of Housing, 1997).

poor community to comply—even if they wanted to. It also, as would be expected, imposes huge delays on development which no poor community could be expected to bear. For example, a critical path analysis of the time required for a so-called fast-tracked upgrading of informal settlements near Johannesburg estimated that it would take nine years if all normal official procedures were to be followed—and this was by the local authority, which was, in almost every respect, the body which would grant such approvals.[15]

Another good example of the procedural difficulties is that of Malaysia: In the mid to late 1980s developers in Malaysia were required to satisfy 55 different steps . . . which could take five to seven years before they could deliver.[16]

RIGHT AND WRONG

Land invasion is, in almost all moral codes, wrong. Is it different in kind from theft of a car? After all, it is theft of property. And is the fact that the land is lying unused sufficient to justify taking it for one's own use? To those who may have had cars stolen when they were not being used, such justification appears, on moral grounds, rather weak.

But while private morality decries appropriation of assets belonging to another, is there an alternative view which looks at the greater good of society? Is there not a duty on society and, by extension, the government to intervene in the private or public market so as to allow the poor to use their own energy and very limited resources to maximum effect?

As we know, this latter view is not new: after the 1976 Habitat Conference, this became part of the mantra of international advice on housing policy. However, there is a second aspect which received less emphasis, and which should, perhaps, be considered in more depth.

It is to be hoped that the average reader who looks at the table of offences and fines would have thought "how absurd." And it *is* absurd that ordinary people should have to comply with these highly technical requirements. But does that mean that such legislation is wrong, or unnecessary? Unfortunately we cannot take the easy way out and say that these regulations are inappropriate or wrong. Anyone who has witnessed the depredations of commercial developers will know what can happen in an unregulated environment. Priceless natural habitat is destroyed, people are given products that are unsafe and unhealthy; excessive prices are charged for defective products; the rights of individuals are infringed by selfish and inconsiderate persons; neighborhoods are polluted by noise and unsanitary conditions. Regulation indeed protects us from a litany of problems.

How then should society respond to this situation? Should there be different rules for different segments of the population? And if so, how do we decide how to fit the rules to the circumstances? Several countries have adopted so-called Grade 2[17] building bylaws. These allow local authorities to relax bylaws designed for conventional development so as to suit the needs of the poor—for example, by permitting the use of semipermanent materials. However, one must ask whether this is the real solution, as the *process* to be followed in order to obtain approvals remains the same, and just as inappropriate.

In many circles there is a certain social cachet in defying the law, for example, avoiding tax, ignoring speed limits, and so on. Something of the same feeling no doubt accompanies the achievements of the residents of informal settlements as they see the fruits of their labor growing in spite of, and in defiance of, established society.

POSITIVE DISCRIMINATION

Our case study shows the potential for a different solution. Communities which are outside the law develop their own law. This law is often applied with more consistency than the law developed by others for their own benefit. Furthermore, within Africa, the man in the street often has a strong sense that much of the formal legal system is a colonial hangover, and therefore has no real legitimacy. How much more rewarding is it to defy such laws than it is to defy the laws which are a product of the society in which one lives? Thus in the rural environments of Africa, not only are there very few rules, but the few that exist are understood and accepted by all. In the unusual event that the rules are defied, justice is administered by the village elders or chief. This is what are commonly labeled cultural standards.

There are examples from all over the world which show that self-regulation can be successfully used as an alternative to the imposed and artificial bylaws and regulations applied in our story.

Does this mean that we should allow the residents of informal settlements to do their own thing, and that formal government should not interfere? The answer is clearly in the negative: in all societies there are people who will exploit others for their own benefit, and whereas in the rural areas they are constrained by the social climate, in the urban areas the scale of the situation and prevailing economic values make such a situation wishful thinking.

International experience suggests that there is a need for regulations, but that they must be such that they facilitate the development of housing that is affordable:

> There is ample evidence that when formal land development parameters (such as minimum plot sizes, setbacks, and infrastructure servicing standards) are not benchmarked against what the local population can afford to pay, most households (not just poor households) are excluded from access to formal land ownership.[18]

Specifically, in relation to informal settlement upgrading:

> Informal settlement upgrading can only be successful if regulations and procedures relating to land tenure, land use and building standards are flexible and appropriate for the needs of residents . . . Appropriate regulations and procedures that . . . allow for some flexibility and for some degree of 'less formality' need to be developed.[19]

These recommendations are clearly pointing toward the need for different regulations for different income groups.

DOUBLE STANDARDS?

The term *double standards* has a pejorative ring. It could be considered discriminatory to have one rule for the rich and another for the poor. Some would say that this approach entrenches class divisions and permits the very same low standards which regulations are supposed to prevent.

Before dealing with this point directly, we should step back to look at the purpose of standards.

> Shelter represents an investment by either individuals or society. Standards must, therefore, reflect the capacity of individuals, societies and their institutions to meet the housing shortage and to enhance the quality of life. In a stratified social system, where a large proportion is below the poverty line, the economic aspects of housing acquire a different meaning and importance. Although standards should obviously help to create a desirable living environment, they must be realistic in distinguishing between what is essential and what is merely desirable, if for no other reason than economic feasibility.[20]

The essential point in this quotation is the need to distinguish between what is desirable and what is essential. Essential might be defined in terms of minimum health (light, ventilation, and hygiene) and safety requirements.

There is therefore a need for a basic minimum set of rules to be applied. It would be nice if we could start from scratch in such a way that if some sections of society feel the need to use "higher" standards, for the protection of their interests, they would be free to do so, thus starting from a basic minimum and going upward, rather than taking a standard developed for the middle class and going down.

Unfortunately, in most countries the existing standards are those designed for the formal sections of the city, which often have colonial antecedents. Anything appropriate for informal settlements must therefore carry the stigma of "low standards." Therefore it will be necessary first to define the "low standards" to be applied, and second to identify where they will apply.

This is not always easy, and will often require major legislation.[21] In Zambia, this was addressed by the designation of the so-called squatter settlements which were to be upgraded as "Improvement Areas" under the Housing (Statutory and Improvement Areas) Act.[22] This gave the local authority the power both to determine what the bylaws would be and to identify the areas where they would apply. In the event they adopted standards which can only be described as completely minimal (see Box 5.7).

Box 5.7 Zambia: building standards adopted by the Lusaka City Council.

The following is the full text of the standards applied to informal areas in Lusaka.

Materials
1. Floors: Floors of rammed earth are to be permitted, provided they are at least 15cm above ground level. Participants are to be encouraged to provide a floor of screeded brunt brick or 50cm concrete when means permit.
2. Walls: The minimum standard for walls shall be sub-dried bricks. Stabilised earth bricks, burnt brick and concrete blocks can also be used.
3. Roof: The roof should be constructed of corrugated iron or asbestos cement roof sheeting in good condition erected on sawn purlins for which purpose rift sawn gum poles are acceptable.

Size
1. The minimum habitable room size should be 6.5 m².
2. The minimum ceiling height should be 2.2 meters, with an average height of 2.4 meters.

Ventilation
1. In addition to any doorways, there should be openings equivalent to 5 per cent of the floor area of each room.
2. Permanent Ventilation: One air brick, 15 x 20 cm or equivalent shall be provided for every 7 m² of floor space.

Structure
1. All houses must be approved as structurally sound and in the case of new buildings must have foundations of at least 22 x 8cm under external and load-bearing walls.
2. All purlins shall be secured into the wall.
3. Suitable lintels must be provided over all openings unless metal frames are used, e.g. around windows or standard metal door frames.

Source: Richard Martin and Robert Ledogar, *A squatter settlement in Lusaka, Zambia "George Compound", Report submitted to the United Nations Centre for Housing Building and Planning* (Lusaka: Lusaka City Council, 1977).

COMMUNITY-BASED LAWMAKING

Developing the standards/bylaws in such a way that they are meaningful and effective is an interesting opportunity for community-based development in action. There are many documented examples of communities making their own rules. One of the most interesting cases from the point of view of settlement management is from Zambia, where residents of an existing settlement decided to develop a new area which would set higher standards. The leadership formed a committee to handle the area. The land was annexed, plots were laid out, and applications were received. Unlike the original area, where land use reflected rural patterns, and houses were placed around common open space, the new development was laid out in straight lines, with clear reserves for future road construction. Everyone who was allocated a plot had to agree that they would build "proper" houses, that is, with concrete blocks. As a result, within only a few months the area looked just like one of the lower middle-class areas to which so many squatters aspired.[23]

There are many such examples. In Kenya, those who joined a water scheme at its inception had to pay a lump sum and contribute labor to help construct the system. Stragglers had to pay higher connection fees. As stakes were raised, members became more involved to guard their contributions.[24]

In Mumbai, the residents of the land on which a squatter settlement called Sanjay Gandhi Nagar had been built had an opportunity to buy it. They immediately made rules regarding the land. One example was that none of the residents could sell their houses for ten years, and if they did then want to sell, the society representing all the residents would have the right to buy the building first.[25]

As described in Chapter 2, in Addis Ababa, people are coming together in groups to form cooperatives, illegally purchasing land from farmers on the city periphery, and "developing" the area. They are careful to meet all the prevailing planning standards—in terms of minimum plot size, road width, right-of-way, and so on—in the hope that they will be legalized sooner or later (see Box 2.6). With each household contributing a fixed sum every month toward capital and maintenance costs, just as in any formal housing association, many of these cooperatives have managed to get the basic trunk infrastructure for electricity and water supply.

The quality of the houses built on these plots varies, depending on individual affordability, but in general all permanent construction is in compliance with city regulations. In other words they are "planned" developments, which people have undertaken in the face of government inaction to address the distorted land and housing markets, and the extremely short supply of housing that is affordable. Still, the regulations deem them as illegal, and they are constantly under threat of demolition from the inspectors.[26]

The preceding examples are of rules which take the form of contracts: "If you want to participate, you will agree to . . ." How much more difficult is it to adopt behavioral rules which apply to all members of a fractured community, and are introduced after that community has been in existence for decades?

The following case shows that while it is difficult to adopt new rules it is not impossible. It concerns an initiative to improve land management in the context of a breakdown of traditional land management practices among pastoral communities in northern Kenya, where there was competition for water and grazing within a context of population growth and insecurity due to bandits.[27]

The population of about 120,000 was divided into twenty-nine neighborhoods each of which elected an environmental management committee (EMC) and each EMC made rules. However, not everyone agreed on the boundaries of the neighborhoods, and different neighborhoods had different rules. Thus a conflict situation developed. The situation was complicated by the lack of legal standing of the EMCs, and the fact that committee members (who were unpaid) felt very reluctant to impose rules on fellow community members. The situation was eventually resolved through efforts of a peace committee, and at the time of the study, the system was working well. Grazing areas were being reserved for sustainability, wildlife poaching had been dramatically reduced, and the EMC's role was being supported actively by the Chief, who supported enforcement of the rules. Inter-ethnic grazing cooperation had been successful in opening up areas previously unused due to insecurity, and allowed formerly hostile groups to inhabit the same area.

This example shows that the community members successfully managed the conflicts which had arisen from differences between neighboring communities, and overcame the problem of weakness in enforcing their own rules.

LAW AS A TOOL OF SOCIAL JUSTICE

A system which permits communities to develop their own rules within a legal framework, and to have an active role in implementing the rules, addresses the need for regulation. By developing the rules themselves they can ensure that they mean something to the members of the community and are written in a form that is easily understood. In other words, they will be in complete contrast to the thousands of pages of legalese verbiage of current acts, regulations and bylaws.

Once more this can be illustrated by the example of Zambia. In "Improvement Areas," the Lusaka City Council was able to introduce much simpler regulations (reproduced in Box 5.6)—simpler to understand and simpler to enforce. These reflected local practice[28] and understanding, so were very easy to enforce. Other matters, which would normally be considered the prerogative of town planners, such as the extension of houses and prevention of encroachment on road reserves and neighboring plots, were dealt with by the community through the branch committee. Not only did this system work without conflict; it also required virtually no administration, either to approve the construction or to enforce in matters of noncompliance.

In Swaziland, where the traditional leaders play an important role in the rural areas and peri-urban areas, we proposed that the development process (including the making of rules regarding what can be done and how it should be managed) should be under the control of the local community, with the Chief as advisor and ultimately the authority by whom rules are enforced.[29]

An administrative legal system has three components: the *law* itself and the rules to be followed under it; the *system* which must be used in order to comply with the law; and the *means* by which the law is enforced. There are many examples of laws being adapted in one sense but using the conventional systems for the other two components. Experience suggests that unless all three aspects are adapted to the needs of the community concerned, the value of the laws will be essentially lost.

There is no doubt that the activities referred to in our case studies are echoed daily in thousands of cities around the globe. Ordinary people are effectively demonstrating that the law is irrelevant, and the authorities trying to enforce such irrelevant laws are proving, yet again, that the law is an ass. If it were not for the fact that they can, and often do, use violence in so doing, the situation could be considered funny.

Many people feel uncomfortable about one law for the rich and one for the poor. However, applying a single set of rules to all members of society does not work, and there is a need to enable innovative, bottom-up rule making. If correctly undertaken, with the input and management of the community, this can enable development to be managed in an effective and user-friendly way. Likewise, if inputs from the community are not adequately sought, even projects with the soundest of intentions can fail miserably.

But we must address the question of first- and second-class standards. There are many who will question this for two reasons. The first is the view typically advanced by politicians, which can be summarized as "how can we condemn people to substandard conditions by using low standards?" The second is that it is unfair on the better-off sections of society who have to comply with tough systems at the cost of time and money. For this section of society it devalues the law that some people should be allowed to "get away with it" while they cannot.

Of course readers will recognize the political view as confusion between standards and products. Of course it is desirable for everyone to have the same desirable standard of housing and urban infrastructure. But in effect standards have merely a normative role.

The second point should be addressed with more attention. As economies grow, the power of capital increases, and it is this very same capital which can be used as a power for good or bad. In the hands of the individual we do not have any problem with, in general, allowing people to establish their own standards provided they meet the needs of health and safety. But allowing the same freedom to developers and landlords has very different implications, and it was the absence of these controls which caused the slums that we referred to in Chapter 1. The same slums continue to be formed even today as a result of the pervasive behavior of landlords—as is the case in Nairobi and other large cities, where illegal and substandard rental markets are thriving—in the absence of the state's ability to enforce proper regulations. The mass production of housing as well as commercial and industrial buildings has implications for the public from which society cannot absolve itself. Regulations and controls are essential for these circumstances, and they will need to be sufficiently sophisticated to ensure structural integrity and sound construction.

Part II

6 Constructive Engagement
Structuring Participation

*The word **participation** is one of the most abused in the English language—abused both deliberately and innocently. This book's central theme is the creative and meaningful involvement of users in decisions that affect them, and by users we include residents of informal settlements whose environment might be affected by development proposals, and all users of public services. Ultimately we are talking about how decisions are made, and by whom.*

The history of the abuse of the word *participation* is about forty years old. In many ways it is odd that, before that, people saw no need for the concept, possibly because the power of the state, and the types of state intervention which took place, and the general setting and context were such that affected parties saw no need (or opportunity) to assert their views. Or maybe the complexity and reach of the modern state requires solutions that are new.

However, it would not be appropriate to leave the impression that the involvement of citizens in government is a new idea. Thomas Jefferson's words are as valid today as they were in 1820:

> I know of no safe depository of the ultimate powers of society, but the people themselves; and if we think them not enlightened enough to exercise their control with a wholesome discretion, the remedy is not to take it from them, but to inform their discretion.[1]

It would be interesting to explore this history, but this is not the right place to do it. This is not supposed to be a scholarly book: its focus is a practical one. Nevertheless, this chapter must address some of the debate in the field of participation and consultation so that we can establish exactly what we want to do, and how the system can and should accommodate the needs of the communities in terms of their involvement in development, the proper allocation of roles of all parties, and the establishment of systems within which all parties can work to their best advantage.

There are practical considerations in participation. For example, there are perceptions that it is slow and expensive. It may also be inconvenient because it might reveal flaws in the government's plans, or it could be the vehicle for the political opposition to garner support. These comments by

South African politicians echo the views of many a local and central government official and politician we have worked with:

> People are tired of telling the government what they want. They want the housing issue to be solved. We have heard the people. Now we must go away and do that.
>
> (Provincial Government spokesperson)

> The people are tired of talking to officials. They consult with them so much, but nothing happens—they want implementation. They are 'over-consulted.'
>
> (Councilor)[2]

This chapter, and the remainder of this book, is designed to provide a framework for the establishment of a sustainable system of development in which the community is the primary decision maker, and to address the skeptics represented by the preceding quotations. But just to reconfirm the point that participation is not an airy-fairy concept that is practiced only by a few way-out people, we should remind ourselves that it is an essential part of the democratic system. There are those who abuse the concept of democracy by claiming that, once elected, they have the right to ignore the views and the interests of the weak and poor: they should reflect on Amartya Sen's views:

> We have reason to value liberty and freedom of expression and action in our lives, and it is not unreasonable for human beings—the social creatures that we are—to value unrestrained participation in political and social activities. Also, informed and unregimented *formation* of our values requires openness of communication and arguments, and political freedoms and civil rights can be central for this process. Furthermore to express publicly what we value and to demand that attention be paid to it, we need free speech and democratic choice.[3]

In a rejoinder to the view that democracy is an expensive luxury, using two very practical examples he points out that, for example, no substantial famine has ever occurred in any independent country with a democratic form of government. Debate, which some construe as conflict, is an important medium in public education, as illustrated in the very controversial issue of birth control. It has been shown that public discussion has an important role to play in reducing the high rates of fertility that characterize many developing countries, as young women absorb information and values from public debate.[4] Allowing participation, and thereby facilitating a process by which people make up their own minds, can therefore act to promote sound decision making. This is in contrast to the popular view that participation allows rabble-rousing demagogues to sway crowds to take irrational and impetuous decisions. Yes, this can happen, but as this book will repeatedly emphasize, the process which is followed is the key.

Turning to participation in action, so to speak, it is useful to refer to the most famous work in the field: Sherry R Arnstein's "Ladder of Citizen

Participation." This establishes a typology of eight levels of participation to help in analysis of what the terms *participation, consultation,* and other activities which purport to involve communities, represent. She chooses to arrange the eight types in a ladder pattern with each rung corresponding to the extent of citizens' power in determining the end product (see Box 6.1).

Box 6.1   Eight rungs on the ladder of citizen participation.

The bottom rungs of the ladder are (1) Manipulation and (2) Therapy. These two rungs describe levels of "non-participation" that have been contrived by some to substitute for genuine participation. Their real objective is not to enable people to participate in planning or conducting programs, but to enable powerholders to "educate" or "cure" the participants.

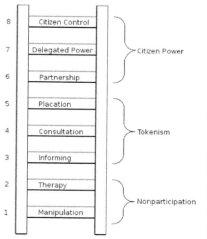

Rungs 3 and 4 progress to levels of "tokenism" that allow the have-nots to hear and to have a voice: (3) Informing and (4) Consultation. When they are proffered by powerholders as the total extent of participation, citizens may indeed hear and be heard. But under these conditions they lack the power to insure that their views will be heeded by the powerful. When participation is restricted to these levels, there is no follow-through, no "muscle," hence no assurance of changing the status quo.

Rung (5) Placation is simply a higher level tokenism because the ground rules allow have-nots to advise, but retain for the powerholders the continued right to decide.

Further up the ladder are levels of citizen power with increasing degrees of decision-making clout. Citizens can enter into a (6) Partnership that enables them to negotiate and engage in trade-offs with traditional power holders. At the topmost rungs, (7) Delegated Power and (8) Citizen Control, have-not citizens obtain the majority of decision-making seats, or full managerial power.

Obviously, the eight-rung ladder is a simplification, but it helps to illustrate the point that so many have missed - that there are significant gradations of citizen participation. Knowing these gradations makes it possible to cut through the hyperbole to understand the increasingly strident demands for participation from the have-nots as well as the gamut of confusing responses from the powerholders.

Source: Sherry R. Arnstein (1969). "A Ladder of Citizen Participation," JAIP, 35, No. 4, July 1969, pp216–224.

In certain respects the context within which this famous article was written is out of date. And we should remember that Arnstein was referring to the United States of the 1960s, when the whole concept of citizen participation had been won by political activism and protest.

It is of interest that she places consultation halfway up the ladder, thus clearly drawing a distinction between it and full participation. However, not everyone agrees with this distinction, and we think it is useful here to go into the concept of consultation in some depth so as be clear about what it is, and its place in development.

CONSULTATION

Consultation is basically the act by a decision maker (typically a government agency) of inviting input in order to gauge the views of the public. How the agency responds to these views is the responsibility of the agency concerned, and once having provided those views, the input of the public is ended.

The point here is that before inviting contributions from the public, the agency must determine what it is on which the public is to be consulted, and how that should be communicated.

Perhaps because so much emphasis has been placed on participation, the literature regarding consultation in third world countries is remarkably thin. An alternative explanation may be that the infrastructure required for systematic consultation has not been established in many developing countries, whereas the machinery for participation is often established outside official channels under the umbrella of specific projects.

At one level, market research is a form of consultation, because consumers are asked to comment about their rating of a product, or their needs in relation to specific aspects of consumption. This can be used by a water utility as much as it can by a shampoo manufacturer. At the other end of the scale, consultation can be used as a factor in the taking of decisions of great national import. For example, the people around London's airport had the right to be, and were, consulted regarding the construction of a fifth terminal. There was a very large number of objections on the basis of concerns about the environment, noise pollution, loss of property values, and so on. The pinnacle of the consultation process was a hearing conducted in accordance to strict rules of natural justice and fair procedure. But, in the end, the decision was taken to proceed, and in many ways the people consulted felt that the whole process had been a mockery.

A similar sense of outrage was felt by a community in South Africa. They used to be within a province which was well run and relatively prosperous, but a decision was made to consider moving the boundaries of the province so that the community would be in a relatively poor province. A process of consultation was used to obtain the views of the residents: they used the opportunity to express total opposition to the change. Only a few days after this process had been concluded the minister used his powers to declare that the provincial boundaries had been changed. The community has since taken the case to the constitutional court on the grounds that their opinions were ignored and the consultation was a sham undertaken without any intention of listening to the opinions obtained thereby.

Consultation also means many different things to many different people, as the following quotation illustrates.

> This planning officer came along to our community forum to tell us about the traffic-calming scheme. He said he'd consulted us about the scheme already. But he hadn't, he hadn't been to any of the tenants' association meetings, we'd seen nothing about it on the estates. All he'd done was display something at the town hall and advertise it in the paper, and he called that consultation. That's not good enough. We sent him away with a flea in his ear, telling him to do it again and do it properly.
>
> —Tenant's representative, consultation forum
> in a South East District Council, UK. [5]

It is interesting to explore the different uses to which consultation is put and to illustrate different modalities, but our main search is for a methodology to evaluate consultation methods.

Schwartz and Deruyterre state that "to consult" has a number of meanings both in the development literature and in ordinary discourse.

- *Informing*—as when a government agency tells a community why, how, and where it plans to install an electric grid to benefit the community, in the manner of city planners.
- *Eliciting opinion*—for example, when planners consult a technical expert for information laypeople do not have, or for advice about how to implement the decisions made by nonexperts.
- *Involvement and dialogue*—This last element is treating consultation as a prelude to or a precondition for effective participation. Such consultation goes beyond simply informing the community of development plans, and even beyond taking community members into account as experts on local conditions and priorities. Community consultation means that the community, planners, and lending agency staff enter a dialogue in which the community's idea and priorities help shape projects.

To sum up, community consultation is a *process* though which a donor or government agency *communicates* with and *informs* communities of its goals and actions.[6]

Box 6.2   The ladder of consultation.

LEVEL 4
To reach **agreement** or partial agreement on a design for some proposed initiative. This is most likely to occur in face-to-face meetings within a supportive and consensual climate.

LEVEL 3
To **exchange information:** that is, a combination of the preceding categories. This offers little chance for stakeholders to understand each other's position.

LEVEL 2
To **get information:** that is, to gather information to be used in decision-making by government etc. and taking pains to tap a sufficient variety of opinion.

LEVEL 1
To **give information:** that is, to inform the community. When the main purpose is to give information it may be appropriate to use the mass media.

Source: Adapted from Community Consultation Checklist, http://www.scu.edu.au/schools/gcm/ar/arp/comcon.html, pp. 2–3.

A useful illustration of the possible different outputs of consultation is provided by the diagram in Box 6.2.

They also provide a good overview of the value and place of consultation in development, and identify three important criteria for ethical policymaking:

- *Complete consideration of impact,* that is, identifying stakeholders who will be affected by development, and a complete listing of the different kinds of impacts (social, cultural, economic, ecological, environmental, political, nutritional, health, etc.).
- *Need for information*—a clear understanding of the need for further information to allow for better means of dealing with unintended consequences.
- *Awareness of alternatives,* that is, in order to minimize adverse impacts, alternative choices and their risks should also be considered. Otherwise it is not possible to weigh the advantages of alternatives.[7]

Community consultation is an important tool for undertaking the preceding, and is a "precondition for the devolution of power and authority to local groups."[8] It is not an end in itself: merely going through the motions of consulting with affected groups can *raise expectations* about local involvement in the development process, and when people realize it is a show, they will become bitter and may well be unwilling to "participate" in future projects. Shah illustrates the problem of tokenism in consultation.

> Consultation fatigue is another constraint . . . The Minister of Finance said that 38 consultations had been done in the past year in the country. They were sick of being consulted: learn from the previous ones and act on it! Doing participation for its own sake is not useful, we must think about the outcomes.[9]

Boxes 6.3 and 6.4 provide examples of consultative and non-consultative approaches.

Box 6.3   Case examples of Mayerema, Guatemala.

Taking a case study of a sawmill project at Mayerema, Guatemala, where existing forests were to be put to commercial use, shows the disastrous consequences of a lack of consultation, and the need to differentiate the different stakeholders and consult with them in different ways. The sawmill owners would typically bring their lawyers to meetings; whereas dealing with semi-literate users of the affected forests--hunters and indigenous groups—required a face-to-face strategy. In between were the general public for whom newspaper articles and radio programs were the most effective communication medium. The point being made here is that there was a total failure to:

- determine whether the concept of the project has been communicated to affected groups
- determine who might be affected
- verify that affected people have been able to voice their concerns, and
- identify appropriate channels of communication with different stakeholders.

Box 6.4   Case example of consultation processes in Ghana.

In Ghana, a very useful comparison has been made in a study by the Council of Scientific and Industrial Research of Ghana between a consultative approach to project implementation in a Water Research Institute (WRI) project and a non-consultative approach in a project undertaken by the Soil Research Institute in Ghana. The former—a project to reduce infestation of the waterways around Accra and Tema by the water hyacinth—adopted a consultative approach and the project was implemented relatively smoothly. The other was a soil and water conservation project. The projects are compared as follows:

"Project Consultation and Formulation Process"
The process for formulating the two projects in the two institutes was different. Whereas the soil and water conservation project was fully developed at the Soil research Institute without consultation with the eventual implementing agency of the results – the Ministry of Food and Agriculture – the water hyacinth project had input from the Environmental Protection Agency and other stakeholders. This gives a clear indication of the need for alliances to be developed and awareness raised, which eventually will be of decisive importance in obtaining approval of the projects. . .

"Public Education and Awareness Raising"
Policies are usually made in order to enhance the well-being of the nation as a whole. However, in most cases the potential benefits may not be obvious or the issue may be too abstract or complex for the public at large. In the case of the water hyacinth, a national public education campaign was launched contributing to the understanding of the problem identified, informing the public about its potential hazards, and the necessity of their co-operation. With the support of the EPA, it was possible to put up posters nation-wide, warning people about the dangers inherent in water hyacinth and pointing out the illegality of growing, transporting or trading in the weed. The soil and water conservation project did not involve the media to any great extent. Although it organized the occasional workshops and field days, it did not have an impact on national policy makers through these media.

"Commitment and Ownership"
From the comparative analysis, it has become clear that underlying the above-mentioned notion of awareness and allegiance, there is another important difference between the two cases: the building of commitment to and ownership of the identified problem. In the water hyacinth case, WRI structurally involved policy makers and practitioners in the process from the very beginning. In doing so, they were made aware of the potential problems, of the need for action at the national and local levels, and of the necessity of funds. At the same time, general public opinion was mobilized to be aware of the potential danger and the need to eradicate the weed."

Source: Council for Scientific and Industrial Research, *Ghana: Agricultural Research and Policy Change in Ghana*. (RAPnet: Case studies. http:www.gdnet.org/rapnet/research/studies/case_studies/Case_Study_25_Full.html)

PARTICIPATION: BACK TO FIRST PRINCIPLES

If, as we have seen, there are effective and ineffective consultative procedures, is it not enough to make sure that the procedures for consultation are thorough and fair? Emphatically not. The point to be made here is that consultation is different from participation, as in consultation the locus of decision making is the "Authority"—the one initiating the consultation. By contrast, the locus of decision making in a participatory project is the affected community itself.

Clearly, the concept of democracy implies a right to participate, but we should recall here that the conventional electoral democracy is not structured in such a way that the public has any right to participate in government per se. This is left to the elected leaders who, once elected, assume an implicit right to make decisions without further consultation. The only justification for consulting the electorate on specific issues—the referendum concept—is that it will give legitimacy to decisions which are potentially divisive.

But referenda are a consultative, not a participative, tool.

It is therefore necessary to go deeper than political systems and look at the realities of the situation today, where people are increasingly frustrated at being marginalized by society at large and urban systems in particular. Two things have changed to bring this about. The first is that the scale of the problem is far greater than it was twenty years ago; and the second is that people are increasingly demanding to be heard. They may have little political power, and even less money, but they do have a voice. Following, we shall look at the relationship between participation (as we use the term here) and the politics of protest and activism. For now, we shall concentrate on the need for, and value of, participation in development projects.

The term *ownership* has been used frequently to justify the need for participation. There are, of course, two senses in which ownership is applied: the first is the metaphorical use of the term: "taking ownership" of decisions, which means seeing decisions as a product of the party that takes ownership. The second is its literal meaning, for example, as when a community develops its own water supply network.

Here, we concentrate on the first question. What is the importance of taking ownership, and what implications does ownership have for commitment and willingness to follow a decision through, under favorable and unfavorable circumstances? After all, are people not rational? Is it not true to say that if I have a good idea, and someone else has a better idea, I will (being a rational being) adopt the better one? If only the humans worked in such a sensible way: unfortunately, it is not that easy.

For example, let us picture a situation in which a community has recognized the need for storm-water drains on their site. They decide on a certain route but it requires that some trees should be cut down. They are aware of that, but see advantages in that the trees can be used as firewood, and consider that it is the most direct and efficient solution. Along comes an environmental group which objects to the destruction of the trees. They propose a different route, which avoids the trees, and even though it is longer, they—the environmentalists—say that funds can be found to meet the difference in cost. Surely, from the rational point of view the latter option is better. So the community is persuaded: they do not have a logical leg to stand on. But yet, the enthusiasm and drive which was behind their original proposal, which they were fired up to implement without delay, evaporates. They no longer have ownership over the idea, and something has happened which no one can fully explain. The project fizzles out.

"A man convinced against his will is of the same opinion still." Robert Burns's famous line sums it up, although it doesn't explain why this should be, and why we tend to lack commitment to implement decisions that we might agree with at the cerebral level, but over which we have no ownership.

One of the most useful and cogent explanations for this phenomenon is the phenomenon of cognitive dissonance.[10]

To illustrate the concept, let us take an example. If, after long and careful consideration, I buy a car, I will take some pride in my decision. I have carefully weighed up the different qualities in the car I want: the trade-off between cost and size; its acceleration and fuel consumption, and its features in terms of central locking, air conditioning, sound system and satellite navigation, and so on; I have also decided on a color. Before buying the car I will also have read advertisements for, and read reviews of, other makes of car. Since these advertisements, and other reviews, may suggest that another make of car is preferable to the one I choose to buy, as soon as I have taken a decision to buy I experience cognitive dissonance when I read such material. To avoid that dissonance, I avoid reading those advertisements and reviews; or if I do, I scorn them as being ill informed or irrelevant to my needs. An experiment showed that people who had just bought a car were very selective in the advertisements that they chose to read (preferring to read those about the car they had just bought), whereas people who had had a car for over two years were generally interested in advertisements for all makes of car.[11]

Once I buy a car, and discover, for example, that a new and improved version has just been put onto the market, I am put into a position of cognitive dissonance because I know I have spent a lot of money, possibly rather hastily, on a product that is not the best on the market for my needs. This knowledge makes me uncomfortable, and is quickly suppressed. I now find ways to reduce the dissonance by identifying features that *my* car has and the new version does not: for example, I will convince myself that my car looks better, or has better seating, or whatever. Alternatively, I will convince myself that I could not have waited a year before the new one came out—this happens all the time. By these arguments with myself I find convincing reasons to justify my decision to buy the car, and thereby remove or reduce the cognitive dissonance.

In experiments with rats, it was found that when placed in positions of dissonance, they started to look for other attractions in the situation they were in. Just as a cat or dog will start to groom itself when placed in embarrassing positions, so too the rats pretended to themselves that they had made the right decision by finding alternative attractions in their situation. Festinger puts it this way:

1. The existence of dissonance, being psychologically uncomfortable, will motivate the person to try to reduce the dissonance and achieve consonance.
2. When dissonance is present, in addition to trying to reduce it, the person will actively avoid situations and information which would likely increase the dissonance.[12]

The impact of this phenomenon on decision making is profound. We shall look at four aspects: dissonance in decision making (i.e., the dissonance which is created by having to choose between two exclusive alternatives); dissonance created by the reversal of decisions; dissonance in group decisions; and the impact of forced compliance. Readers will recognize that all of these instances feature in the implementation of any participatory development.

Dissonance in Decision Making

Many decisions are difficult, and the more difficult they are, the more conflict is created in the mind as the person tries to reconcile the attractions of each one. At this point we should make clear the distinction between conflict and dissonance. The conflict arises when the decision is being taken; dissonance is produced by the knowledge that the person cannot adopt both solutions: a choice has been made. Once the decision is taken, therefore, in order to reduce the dissonance created by this difficult process the decider will concentrate on the relative unattractiveness of the rejected alternative in the hope of reducing the cognitive dissonance. At the same time, he or she will attempt to enhance the value of the decision by concentrating on the good points of the choice made.

Reversal of Decisions

Once a decision has been taken we are naturally very reluctant to reverse it, because that implies that the original decision was not considered with sufficient care or understanding. To be forced to reverse a decision therefore creates a very dissonant situation. In an experimental situation, therefore, those who took a decision with confidence were very reluctant to reverse it, whereas those who had difficulty taking a decision were more easily persuaded to reverse it. Either way, considerable dissonance is experienced by those who change their decision, which has to be reduced, to some extent, by holding on to the original opinion as the right one and only making a show of agreeing to reverse the decision.

Dissonance in Group Decisions

The same is true to a much greater extent when people are working in groups. Where someone has publicly declared themselves to hold a particular view in front of a group, they feel a much stronger sense of commitment to that decision than they do if they have kept their views private.[13] It therefore should not surprise us that if that person is then required by forced compliance to change that course of action or decision this causes considerable cognitive dissonance. In such circumstances it is normal for public compliance and private opinions to differ.

Impact of Forced Compliance

How can compliance be forced?—either by reward or punishment. In such situations the magnitude of the reward or punishment in relation to the seriousness of the decision will greatly affect the degree of dissonance. So, to use Festinger's example,[14] if you are offered a million dollars to publicly state that you like reading comic books (whereas, in fact, you hate comic books), the degree of dissonance will be very small; whereas, to go from the ridiculous to the serious, were you to offer the same amount to a devout Hindu to publicly endorse McDonald's burgers (which are, of course, beef), and he did—then the dissonance would be very substantial. Clearly, having made the statement, his opinion about the rights and wrongs of eating beef would not be changed, but he would nevertheless face the dissonance of having made a statement in public that he could not, in private, ever agree with. Therefore, there is a relationship between the importance of the opinion involved and the magnitude of the reward or punishment necessary to elicit forced compliance, and the greater is the magnitude of the dissonance created.

We have been describing the difficulties inherent in changing people's minds, and suggested that this is due to the phenomenon of cognitive dissonance. If the urge to avoid a state of cognitive dissonance makes people reluctant to change their mind, how is it that people do, in fact, change their mind, and some of the most strongly held opinions can be changed from one extreme to the other? Under what circumstances does the value of the personal opinion become subsumed by something else which allows one, without dissonance, to change one's mind?

Here we revert to the importance of the group, and in particular to the concept of leadership. People often accept leadership because they want to avoid dissonance by relying on the opinion of others. By being in a conforming group they have little need to form individual opinions and have the comfort of knowing that the values which they espouse are shared by the group as a whole. The interesting point here is that people are willing to forego their own opinions in order to benefit from the reassurance and comfort that membership of the group provides.

The most extreme examples of this behavior come from messianic cults and the like. People who hitherto have no expectation that the world will end, or that a particular person will guide them, for example, surrender their individuality for the sake of conformity to the values and beliefs of the group. They embrace the group as a means of finding answers, and thereby avoiding the dissonance which they are experiencing in their daily lives. We would suggest that the certainty offered by such groups becomes more attractive as the circumstances of the individual create more dissonance.[15] The fact is that we tend to congregate with people whose opinions are the same as ours; and insofar as they are not, we prefer, in general, to avoid raising the issues so as to avoid conflict.

Good leaders use this group psychology to their benefit. If we, as individuals, like to be with like-minded people, the first thing leaders must do is to reassure the led that they think and feel like them. Once this rapport has been established—and this, we would suggest, can be from a matter of hours to a matter of months—they take the opinions of the people whom they lead into new directions. The sense of their message will therefore be: "I am like you. I think like you, and I act like you. But I have been thinking and I have realized that something is wrong, and must be changed . . ." (and so on). In this way they get people to surrender their minds to them.

This, of course, is the opposite of forced compliance, which, either by threats or rewards, gets people to do something that they truly do not want to do. And if they do, in the end, take this action, in order to reduce the cognitive dissonance, they may harbor an underlying grudge because their behavior or decision was changed.

These phenomena explain so much about why participatory projects can succeed or fail that we have dwelt on them at some length. They help us to understand the dynamics of decision making that can make the difference between the extremes of enthusiastic participation and full ownership of a process, on the one hand, and hostility and resentment, on the other hand.

In a subsequent chapter we shall be discussing how to generate and support a participatory system; how to offer technical support and advice, and how to deal with conflicts and resistance. Here, our aim is, having presented a theoretical explanation for the need for participation, we need to describe what participation means in practice, and how it can be dressed up in a number of interestingly different clothes.

PARTICIPATION IN PRACTICE

Definitions can be tools for limitation, so if participation is defined too tightly its scope could be limited by such a definition. But it is possible to give some indicators as to what a participatory project, in relation to informal settlement projects, implies and offers.

- All decisions affecting the settlement will be taken by the residents.
- The residents will have enough information in order to take that decision.
- The residents will be able to use technical assistance and support in order to obtain the information they need to make an informed decision.
- The project will be managed in a way that is endorsed by the community.

This list may raise many problems in the mind of the reader: for example, how can the community possibly be allowed to take technical decisions; how can consensus be reached on decisions that may adversely affect some people while benefiting others (thus creating a situation of winners and losers); how can poor people be expected to devote their time to such specialized duties; how can uneducated people be expected to understand the technicalities of civil engineering, and so on?

These are all important questions, which we return to later in the book. The bullet points above are more of a target than a minimum requirement, but we feel that they represent a target that is achievable.

It is interesting to draw a parallel between participatory projects and sport—let's say soccer (football). The team consists of eleven individuals, each of whom has individual objectives and career ambitions. During a game, none of them can tell any other member of the team how to play— there is no time for that. However, each of them relies on the others for success: even the greatest player will expect to be passed the ball, and will rely on teamwork in order to outwit the opposing team. The team must work together in defense and attack, and can only win as a team. The captain may have a role in encouraging and inspiring the team, but cannot control them. A bad team is one in which each member has his own strategy: a good team is one which works as a unit. The really important part of the parallel is that, during a match, the team has only itself to rely on. They cannot stop and ask for advice, nor decide to play the match another day. They have 100 percent of the responsibility (much to the chagrin of coaches the world over, who always get the blame if things go wrong) for winning.

If we can, within development projects, establish a relationship between the authorities and the settlement where the residents are the football team and the authorities are the coaches and management, then we are getting somewhere.

To continue the analogy: a good team practices, and practice involves— more than anything else—understanding each player's strengths and weaknesses. It is essentially a collaborative effort, and the more frequently a team plays together and practices together, the more likely they are to succeed. They can learn movements and cunning tricks which can be communicated by little more than the rise of an eyebrow. But they need help to do this—the coach, who stands on the sidelines during matches and during practice, sees the team as a whole and helps them to work together and improve their game. It is he who helps them to practice passes and movements that will outwit the other side, and it is he who helps them to set their goals.

In just the same way, technical advisors should be available to help the community to work together, to understand the strategies that can be applied, and to help them coordinate their efforts. They do not control the community: they support it.

Unfortunately, it has, as we noted before, become a cliché to sing the praises of participation as a way of obtaining metaphorical ownership of projects. This ownership is claimed to lead to improved commitment by the residents toward maintenance and better performance in terms of cost recovery. In brief, it makes the project more sustainable.

We speak of this in a somewhat skeptical way because as much as successes are claimed for participation in such terms, so too are failures laid at the door of participation. Both may be exaggerated, but the point must be made here that the term is misused, so not all so-called participatory projects are that in fact.

However, we do endorse the view that participation, as described here, will not only lead to better solutions but to more sustainable solutions, because the development will be tailored to the needs and means of the residents concerned. For this reason the community will be prouder of the development and look after the facilities better, *provided that* the project is properly structured—and this is not, we would be the first to admit, always easy.

UNDERSTANDING PARTICIPATORY RELATIONSHIPS[16]

The essence of participation is engagement by the affected people, but it is useful to explore the concept through different perspectives: war, negotiation, politics, and public relations as participation. The reader may raise an eyebrow at what appears at first to be a cynical or even superficial view of a subject which deserves serious treatment. However, there are, within comparatively strange contexts, opportunities to understand the intricacies and subtleties in what is a complex power relationship. The nature of these relationships cannot be described in terms of a one-dimensional phenomenon: there are many layers and possibilities which can best be characterized by painting pictures of specific situations in which elements of participation are present. These elements, like scales of a fish, add to the broad picture. To risk belaboring the metaphor, the scales are components of the whole fish, but are not the fish. They are aspects, superficial maybe, which are nevertheless worth examining.

War as Participation

Let us consider, for example, a community which has been in existence for twenty years. Conditions are bad—a few standpipes constitute the only services. Access is by dirt roads which become impassable after heavy rain. Drainage is bad and many people suffer from malaria. Pit latrines overflow. In spite of this, the community values the place: it is very central and they can live virtually free. The land is owned by the national railways, which had bought it for a goods yard, but since then they have constructed a large goods yard on the edge of the city where road access is better. The value of the land meanwhile has escalated and the railways decide to sell the land for development.

The community has no history of activism: it has been every person for him/herself; a pleasant atmosphere of live and let live has prevailed. There is a strong sense of mutual support within the community: they look after each other in times of hardship and sickness, but have not been active in the field of community mobilization. There has been, until now, no need.

When the sale of the land is announced, the press picks up the story and sends camera crews to the settlement. Until then the community has not realized the seriousness of their situation. Something now has to be done: if they don't act now, and act strongly, they will lose their land and houses.

Overnight the community is transformed from a passive to an active one. Neighbors meet to discuss strategies, public meetings are called, and leaders are elected. Many people sacrifice their time to mobilize support from the community. They are now emerging from their previous state of peaceful apathy of individuals into an active community fighting for its very right to exist. They are now showing symptoms of full participation in their own affairs.

The leadership is elected, but can not take its role for granted. Theories about what will work best abound and have to be debated. These debates take place in homes, on the football field, in the youth and women's groups, in the church and mosque. They take place in formal leadership committees and in public meetings.

The choices which are debated include legal action, mass protest, negotiation, and passive resistance. Ultimately these are all strategies to fight a war, and even though this war has neither bombs nor bullets, it is real. The community is literally fighting for its survival. They know that if they lose the battle to remain on their land they will be ejected onto some peripheral site too far from the income-earning opportunities that most of the people have in their present location for them to survive. Not only will the community as they know it be lost—so too will their sources of income.

We do not need here to know what strategy is adopted, or how successful it is. But we can, for example, speculate that the community's determination and energy, which is a product of the war which they have declared, allows them to negotiate an excellent deal with either the railways or the new owners. Without war on the railway company having been declared, they would have had neither the leadership nor the community solidarity to have succeeded. War was a trigger for dedicated and effective participation, and mobilized a community spirit which had hitherto been lacking.

Public Relations as Participation

It is truly amazing that so much money and skill is used in the cause of public relations, in the belief that by so doing the public will want the same things as those who send the messages.

If we rewrite the preceding scenario, we can illustrate the point. *Before* actually selling the land the railway company might have realized that they would have a problem with the residents. In the boardroom they would have discussed the best way to get rid of the residents peacefully. Their goal would be to first clear the site, and then sell it, thereby getting a much better price. Some of the additional yield on the sale can be used to buy some land on the edge of the city for the resettlement of the people. Basic services such as gravel roads and water standpipes can be provided.

One must assume that occupants of boardrooms are intelligent people, but their experience of slums and informal settlements will normally have been through the odd glimpse from the car window, horror stories in the press, and possibly buying something from a poor person on the pavement who is, presumably, the resident of such a place. To them, therefore, it is the bad physical conditions which will be the main concern of the residents.

The campaign is therefore a simple one. Just inform the people that they will be given a new site on which to live, and that it will be much nicer than the one they occupy presently. The public relations campaign is, of course, as much looking over its shoulder at the reaction of the press as it is at the community members.

The foundation of the campaign is a simple one: tell the people what we—the caring and socially responsible railways—are doing for them.

A public relations company is engaged, and a carefully crafted campaign is presented at the board meeting. This features massive billboards at the new site, press releases, a local radio campaign in which railways spokespeople would explain the plan, and pamphlets in the local language which would be distributed house to house. A phone number is provided for anyone who wants to ask questions. The phone is operated by the public relations company. The operators are carefully trained to follow the exact wording of the standard answers to the questions which they will be asked.

This campaign is expensive, but the board congratulate themselves in the knowledge that the residents are fully informed and involved. They know that they have done everything they can for the residents—indeed, they are opening the door for a new and better life for them.

An equal and probably more expensive part of the public relations campaign is to inform the public at large what a good job the railway company is doing: the great lengths it is going to in order to help the poor, and the rosy future that the residents of an erstwhile slum will have when they are moved to new and sanitary accommodations. The campaign will run feature articles in the press, glossy advertisements, television interviews, and so on.

The irony of most such public relations campaigns is that the people operating them do not realize that there is a difference between being told what is going to happen to you and wanting it to happen.

So what happens next in our scenario? The leaflets are distributed; the call centre is set up; all the expensive ads are run in the press and radio. And then? As likely as not the residents will say "we were not consulted" and will, without any sense of irony, protest accordingly. The railway company will immediately rush to its own defense: "They were allowed to participate freely in the whole process," they will complain, with a genuine sense of grievance (thinking of the call center), "and now they turn round and pretend that we have done nothing for them."

Maybe the people do and say nothing, but when the time comes for resettlement, they show their resentment by passive resistance if not outright protests.

Politics as Participation

To consider participation as politics, we need first to paint a picture of the formal political system prevailing in the scenario.

Every four years there are local government elections. Even though the settlement is not formally approved in terms of housing and planning, the residents have the vote. Local government elections do not generate much interest—only about 40 percent of the electorate bother to vote, but party politics are quite lively. The party running the local government is the opposition in the national government.

The councilor elected by the community does not come from the community itself. He is an upcoming firebrand in the party who has ambitions to be mayor, or even a member of parliament. His election campaign was bitterly fought between him and a man from within the community. The latter appealed to the older people, but many thought him backward in relationship to the winner.

Under the local government law, the councilor has the duty to establish a ward development committee. On it are supposed to be representatives of special interest groups such as small businesses, youth, women, and so on. Soon after he is elected, the councilor invites people he can trust to a special meeting at which the ward committee is elected.

That was two years ago, and the ward committee has met five times. The meetings are dominated by the councilor. He calls the members "you people" and lectures them on issues which have been discussed by the council. So far, these have been the need for people to register their children at schools; the need for people to keep their area clean (the mayor has an annual clean-up day); the need for mothers to take their children for vaccination; the need for the houses near the road to be painted, as the president is to visit shortly. He also has ideas for special projects which will advance his career by giving him a higher profile. The latest idea is to create a football team in the community.

The meetings are run with a great deal of attention to protocol: reading and correcting the minutes of the last meeting, matters arising, any other business, and so on. Discussions are started by the councilor himself so that everyone knows what they are expected to say. The members find many ingenious ways of agreeing with him while giving the impression that they are making a new contribution.

When the railway company conceives their plan to resettle the residents, they approach the local government and are put in touch with the councilor. He is invited to a grand luncheon and shown much respect. After the lunch, the board chairman takes him to one side and asks his opinion about how the people would respond to a resettlement plan. It is made clear that the company is asking the advice of the councilor—very informally, of course.

The councilor is flattered by the attention he receives, and likes to give the impression that, as a leader, he knows how "his people" will respond favorably. Moreover, he is sure that whatever commitment he makes on their behalf, he can persuade them to do. He knows how much the residents respect him and that they will follow his lead.

The chairman outlines the plan in a general way, and asks for the councilor's views. The councilor thinks it is a good idea: he also wants to take credit for upgrading the standard of living of the residents. This could give him a high public profile—just what he has been hoping for. The chairman asks about consulting the people directly: Should they meet the local leaders? Or even have a public meeting? The councilor rejects the idea firmly. He explains that this will complicate matters and allow rabble rousers (his term for the opposition party) to disrupt proceedings. He insists that he knows what people want and will be able to speak on their behalf.

As the company develops the idea, they consult the councilor on a regular basis. What size plots will people accept? What type of roads and water supply? Will they expect electricity? What about schooling? He gives his input on these and many other subjects. He decides that it might be appropriate to discuss the matter with his ward committee.

Emphasizing that matters are very confidential at this stage, and not really saying anything more than posing the question about what people would want if they were given their own land, he asks their opinion on the question of plot size, servicing levels, and similar matters. Thus armed with the views of his people, he fights hard for the standards to be high, not wanting to be associated with a "substandard" project. In meetings with the railway, he insists that his people will not accept anything less than 200 m² plots. At the next meeting he says they will definitely require waterborne sanitation. This is followed by a demand for tarred roads and street lighting. As these demands are notched up, the railway company begins to get nervous—the price will be far higher than they expected. They ask the councilor to find other money, and after much haggling with the mayor and the chief executive of the local government, he manages to obtain funds for purchase of the land. "The rest," he says emphatically, "will have to be paid for by yourselves."

Eventually, the plans are finalized. The railway company and the city sign an elaborate memorandum of agreement about what each party will pay and do. Now that the deal is in the bag, who will make the announcement? The councilor makes it clear that this is his job, and eventually both the chairman of the board and the mayor relinquish their demand to announce the good news.

Is it necessary to take this scenario further? All the parties think that participation has been undertaken except for the subject matter of the case—the residents. Their perception, quite correctly, is that there has been no participation. Even the members of the ward committee hurry to distance themselves from the decisions by saying that, while the councilor asked them about standards, he did not consult them on the move, the selection of the site, or the costs. Because what everyone had overlooked, and only the railway company really understood—so they buried it in the small print—was that the residents would have to pay the for services on the new land. They proposed a monthly charge which would be affordable to all residents, and would recover the costs within ten years. "Otherwise," said the chairman, "how can we possibly justify the expenditure?"

This councilor may look either naïve or dishonest in this account, but these descriptions have been taken from real life—indeed, too many real-life occurrences to mention. He considers that he is genuinely doing his democratic duty, and would be highly insulted if one was to suggest otherwise. He sees himself as having been elected to take decisions, and that consultation—let alone meaningful participation in decision making—would, in a way, mean that he could not do his job properly.

Perhaps we should continue the scenario. The people reject the proposal totally. Not only do they not want to go and live on the edge of the city; they consider the high standards nice but not worth paying that much for. They have other priorities for their very limited means. Thus, not only does the councilor look foolish; the railway has invested much time and money on a project which is fundamentally unsustainable.

There is a very real tension between the concepts of good leadership and the much-misused terms democracy and participation. The risk is that these concepts will be used interchangeably. In the context of this book it might only be necessary to point out that formal democratic institutions are no substitute for participation, and that forceful leadership must not be confused with good leadership.

Negotiation as Participation

The act of negotiating infers a level of equality between the parties. If they did not have at least some sort of parity, negotiation would not be required, as the stronger party would simply take the decision. In this sense, therefore, negotiation implies participation.

Before proceeding with this line of thought we should step back and create a scenario around the negotiating table. We must assume that both parties have some sort of mandate regarding the acceptable solution, and will—as the jargon puts it—consult their principals in the event that new offers are made which might require the reconsideration of a prior position. Negotiation is therefore a test of leadership, in that the negotiators and their principals must have the support of those for whom they speak, whether it is a trade union, a community, or a country. Ignoring, for now, the test that this can pose for a leadership (as the subject is discussed in more detail next), we can look at the act of negotiating, and how it overlaps with the concept of participation.

Reverting to the case discussed previously, let us assume that the railway company goes to the leadership of the affected community (and assuming that there is a well recognized community-based leadership) and put it to them that they want them to leave the site, and will offer them an alternative.

There are many issues to be discussed in such negotiations such as the location of the site, what the affected people will pay, what the company will pay, compensation for existing structures, and so on. All these matters must be discussed, and the parties have the opportunity to decide for themselves what they are willing to accept. If the parties are negotiating freely, they can never, after this process, claim that they were not involved in the decision, or did not know what was being discussed. To this extent there is full participation.

However, there is a substantial potential qualification to this statement—are the parties actually negotiating freely, or are the residents actually negotiating under the shadow of a gun? If so, they may not, in fact, be participating in such a way that the decisions taken in the negotiations are the considered and final choices of the residents.

The following chart illustrates the positions which the two sides might be starting from in a negotiation.

Figure 6.1   Negotiations between residents and authorities.

Residents' view	Authorities' view
We are treated like outcasts and given no services.	They invade the land so have no rights.
We have invested our own time and money in building houses.	Their shacks disfigure the town and should be demolished.
We are hard working: without us, businesses in this town would have no workers.	They are lazy and are a drain on the economy.
We sell low-cost goods to all the workers in the city.	These street sellers are a menace: they make our city look messy.
We provide cheap services like carpentry, car repair which are otherwise not available.	They are taking the jobs of honest shopkeepers and workshops.
We know everyone in the community and have no crime.	They are criminals: crime has increased since that place was built.
We only want improved services, especially water and roads.	They cannot afford to pay for the town's services, and should go back where they came from.

Typically, inexperienced facilitators will allow the discussions to focus on who is right and who is wrong in their assertions and beliefs. This could be valuable in terms of some factual aspects but is unlikely to change entrenched positions, and if it does so, such changes will be seen as a loss of face. Instead, the negotiator must help the parties to set a mutual goal, and identify steps to be taken to meet that goal. The statement of unilateral hard-line positions must be actively discouraged, and especially doing so in the press or other media. Everything should be arranged so that the parties have the maximum flexibility to respond to the needs of the other. This is why negotiation usually results in a better outcome than court action, which, by its very nature, forces both parties to establish a position which is either right or wrong.

The second aspect of negotiation which is crucial is the degree of freedom which the negotiators have. Effective negotiations cannot take place where each party has to refer decisions back to their principals—thus, in effect, requiring them to both negotiate with the other side and to negotiate with their principals. It is crucial, in the interhuman dynamics that the negotiators have, and are perceived to have, for the authority to make commitments. This is not to say, of course, that the device of "referring this one to my principals" cannot in itself be a useful bargaining tool. "They will never accept this" being a prelude to a break in negotiations to allow both parties to reflect on their choices.

In the scenario shown in Figure 6.2, negotiations might start with discussions about the interests of the parties, put in general terms.

Figure 6.2 Negotiations in general terms.

Authorities	Residents
Objective: *To get the land they need for development*	**Objective:** *To have a convenient place to live*
We need you to vacate this site as it has been sold for development.	What are you willing to do for us?
We are willing to resettle you on our site at the edge of the city.	We are willing to cooperate—and glad that you are willing to spend money to solve the problem.
We welcome your cooperation – as long as you understand that we don't want to see you suffer, but we need to get hold of this land. There is a lot of money at stake.	We understand, but you cannot expect us to make all the sacrifices—the land on the edge of the city is unsuitable. We cannot afford the transport costs to get to work, and it would cost us a lot to build new houses.
We can help you with the cost of building new houses, and can talk about a different site.	Now we are making progress. But what is the problem with us staying here if we pay you some land rent?
You will never be able to pay as much as we shall get from a developer. Also, your houses are just shacks and are a disgrace to the city, and give it a bad name.	If we re-built our houses so that they looked good, then you would have nothing to complain about.
That's not the point—we need you to move.	Some families would like to move to the site outside the city. Can the ones who want to stay move to one part of the site and leave the rest for you and your developer?
In principle we must reject this proposal, as we need the whole site.	So do we. What will your reputation look like if you have to move us by force, and people die in riots? We have no mandate to give in on this one, and moving us will be slow, expensive and very damaging to you.
If we help you to rebuild, you must realize that those who stay will have to make sacrifices themselves, and will need to pay for the land and services.	We have already said we understand this, and will be able to pay the costs, provided that we get the support you offered to help us build new houses.
This sort of arrangement will only work if you accept that the new houses must be suitable for a modern city. No more shacks.	We agree to this, provided that we have an input into the designs and can participate in the building process.
We cannot accept this principle unless we have a full account of how many want to stay, and what sort of housing will be built. This includes the designs and type of structures.	We have undertaken studies which show that only about half of the present families are insisting on staying.

(continued)

Figure 6.2 (continued)

Authorities	Residents
Objective: To get the land they need for development	Objective: To have a convenient place to live
We have come up with housing designs to see how much space they will take up. They show that if everyone lives in four storey apartment blocks you will only need one quarter of the site. This solution is acceptable to us.	These designs are not suitable for our needs. Many families have animals, and small children cannot live upstairs. Anyway, too much space in your scheme is given to roads.
We have come up with high density court-yard housing designs. This occupies about one third of the site. We can only afford this if the people who are moving to a new site do not receive any assistance with their housing.	We can participate in construction of the new houses. All you have to do is to construct the outside walls and we will construct the rest. In this way the appearance will be uniform, and we can build the housing incrementally.
We can accept this provided that basic standards of health and hygiene are adhered to in the new development.	This is accepted, provided there is no rule preventing us from renting accommodation to help us meet the costs of building new housing.
Agreed.	

This simplified negotiation scenario could be written in a different and confrontational way, with threats and counterthreats peppering the dialog. If so, we can be sure that not only would the outcome have been less satisfactory, but that it would have been slower. But the main point of the scenario is to demonstrate the degree of participation in the ideas behind the settlement, and the fact that the residents managed to negotiate a role for themselves in both the design and construction of the new houses.

Before we go into this in detail, it is useful to attempt to answer the question why it is important for people to participate. And to those who maintain that this is a silly question, we should perhaps recall the bygone centuries when the answer to such a question would have been answered quite simply by "It is not important; they should do as they are told." And recall, equally, that there are many states today where such concepts are considered inappropriate or at best superfluous.

Activism as Participation

Now we come to the role of protest and activism. By this we have in mind the situation where communities adopt a stance of protest which is typically characterized by resistance to the authorities and everything that they stand for.

Protest has a role in every society. Often it is suppressed, sometimes violently, and many inhabitants of informal settlements have had many bad experiences of that. Violent suppression of protest is, as we know, a common response from the authorities to an uncomfortable situation. After all, if you are in a position of authority, the presence of informal settlements is, in a sense, a threat to law and order. They invade land, or use it for an unauthorized purpose; they pollute the water and create garbage without paying taxes; they tarnish the good image of the city with their shoddy buildings; they overload the public services of health and education; and then they protest when they think you are not helping them enough.

Such protests are triggers which can be used to justify oppressive action such as demolition and even imprisonment.

On the other hand, it is only natural for people who are neglected and even discriminated against to feel the need for action. Demonstrations are a way of relieving the pressure even if they achieve little, and sometimes it can be satisfying to have a physical confrontation with the police as a means of venting anger.

Jockin Arputham, the founder of Slum Dwellers International, tells a story about how he got the attention of the authorities to the garbage problem in the slums of Mumbai.[17] He asked all the schoolchildren to collect garbage in plastic bags from the settlement in which they were living, and to bring it with them to school. The pile of garbage was so high that neither the teachers nor the students could get into the school. This not only made the point that people can help to clear up garbage, but also that nothing is being done by the authorities to remove it. Strategic use of the media meant that the incident received massive coverage and the authorities were shamed into action.

Another example involves a community which challenged the South African government's decision to evict them. Irene Grootboom and 900 others had been evicted from their informal settlement situated on private land earmarked for formal low-cost housing. They applied to the high court for an order requiring the government to provide them with adequate basic shelter or housing until they obtained permanent accommodation. The government appealed to the constitutional court, which upheld the case of the applicants. It was held that the state was obliged to take positive action to meet the needs of those living in extreme conditions of poverty, homelessness, or intolerable housing. The case was viewed as a victory for the rights of the poor. Ironically, Irene Grootboom herself never received a decent house—either on a temporary or permanent basis—and died penniless in a shack.

This type of direct action can be very effective and is certainly and obviously participatory. Citizen participation in Europe and the United States often has many of the same ingredients: developing a lobby group; developing a strong legal and technical case; demonstrating in the streets and on the site; taking the matter to court.

The famous Greenham Common antinuclear-warfare demonstration which went on for years around a U.S. air base in Britain is one of the most famous of such direct action campaigns: globally there are many thousands of them annually, often focusing on environmental issues and animal rights.

This is a type of participation which has not traveled well. By this we mean that the political and social climate which is necessary for it to work does not exist in most developing countries. The authorities in developing countries take less kindly to being challenged, and results tend to be disappointing. We have been working with a community in Accra called Old Fadama, or Sodom and Gomorrah (seriously—it is so called on an official map of Accra). This is truly an unpleasant place to live, located on the edge of a semistagnant lagoon, on soggy ground that floods frequently. The area was squatted in 1992, and about two years later the Accra Metropolitan Assembly declared it illegal and started to demolish it. The residents obtained a stay of execution, but when the matter of their right to occupy that land went to trial, they lost. In this they were encouraged and supported by a legal NGO. During the prolonged legal process, community expectations were high, only to be dashed. As a strategy, one might ask whether, in retrospect, the community's position might have been better if they had engaged in a less confrontational approach.

Similar support to fight through the courts has been given in South Africa by NGOs such as the Wits Law Clinic and the Legal Resources Centre.[18] However, litigation has two consequences which must be seriously considered before starting on it. The first is, of course, delay. The second is that even if the informal settlers win, the authorities have innumerable ways of dragging their feet or harassing the residents in retaliation.

So the direct action mode of participation, especially if it can be construed as threatening, is not effective in many situations.

Interestingly enough, exactly the same argument applies to the authorities. If they want to take action against a settlement, for example, to remove the houses because they are in the way of a major new road, or too close to a railway track, they have two choices. Either they can rely on their rights in terms of planning or land legislation and demolish, or they can approach the residents and discuss the problem with them. Typically, they will do the former: it seems much simpler and avoids the messy business of compromise and delays. One particular story in Lagos illustrates this very well (see Box 6.5). After being 'resettled' by the government twice, when they were asked to move a third time, they adopted a confrontational approach, ready to go to war with the powers that be.

This sets up the resistance mode: people will resort to violence and possibly court action. They will use the media to publicize hardship stories, and in the long run the authorities will find that demolition without consultation and participation is by no means the quick and easy solution. The conflicts

Box 6.5 Forced compliance: An outcome of unconditional power in Lagos, Nigeria.

Originally based in Maroco this community—some 10,000 strong—was first moved in 1960 from what is now the area between Victoria Island and Ikoyi to a swampy area, which is now the Victoria Island Annex. They lived there for about 30 years, until 1990, when they were told by the government to vacate the area. An announcement was made on the radio, giving them a notice of 7 days to vacate the land. On the 7th day, they were bulldozed and forced out of their homes.

A smaller group of them eventually moved to Ilasan Estate in Lekki corridor, where they were allocated unfinished (previously-abandoned) units. They have allocation letters showing their legitimate occupation of those apartments. However, now, after living there for 15 years, the government has come back to tell them to leave once again—for the 3rd time. There are allegations from the community that "some people have been blocking their drainage networks so as to flood the area and force the people out, on the pretext that that this is flood prone land, which is clearly not the case."

The proposed developer of Ilhasan Estate is LBIC (Lagos Building and Investment Corporation), a lending institution, with the government being the majority shareholder. Below are some remarks from a senior official from LBIC about this community and the estate:

> This is prime property, and this type of [poor quality] living does not befit that place; the land will be redeveloped, whether they like it or not . . . and they will have to move . . . ; Granted that they were allocated those units by government, but the fact remains that the land belongs to the government; of course we will compensate them for what their apartments are worth, so they can't say we did not give them any options. . . .

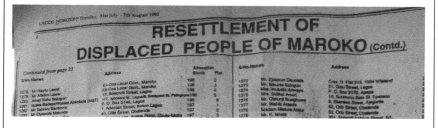

Source: Fieldwork by Ashna Mathema in 2008.

around the encroachment of houses onto the railway reserve in Mumbai and a similar case in Kibera, Nairobi are often cited, with good reason, as an example of participation succeeding where confrontation and force did not.[19] In both cases the initiatives came from the residents themselves, who went to the railway authorities and said that, instead of protesting against them, they would prefer to talk to them. In the end both sides were surprised by the willingness of both sides to be flexible, and comparatively sustainable solutions were found to what had seemed an insoluble problem. It was, indeed, a win-win situation for all.

Protest therefore has a role: that is undeniable, but it should be used strategically, to get attention but not to threaten. With media support it can force a response which good behavior will not. But that response should then be followed up with strategic collaboration to find solutions, and thereby diminish the cycle of demand and counterdemand. The experience has been—and this has been validated by the Slum Dwellers International (SDI) working in many different countries—that engagement is better than confrontation.

Free Labor as Participation

Participation is used in another sense, and we cannot leave the subject without referring to it.

"Oh, yes," we have been reassured on numerous occasions; "the people participated fully in the project. They provided the labor free of charge."

Whether it is building one's own house, digging water trenches, constructing roads and constructing market sheds, this type of participation is not what we are advocating. Of course, if people choose to use their own labor to further the interests of the community, that is fine. They will gain from a trade-off between the saving in cost and the loss of their own time and will have the use of improved amenities which will outlast the personal inconvenience and even hardship which those participating in the labor may have experienced. But that is fundamentally an economic decision, and says little about how decisions are made and who makes them. All too often, this participation is undertaken as a reluctant price to pay for development and has nothing to do with control over what is provided and how. A sad fact that must be faced in such projects is that the opportunity cost of so-called free labor may simply be too high, because if they are to participate in such activities they have to forgo other income-earning opportunities.

PARTICIPATION AS A PARTNERSHIP

The term *participation* assumes a process of decision making. It suggests a situation in which one party—the party with prima facie power—chooses to allow another party to share (= participate) in the process by which decisions are made.

The central theme of this book is that unless there is a transfer of power, participation cannot succeed. How that transfer is made is crucial, and sets the tone for the whole relationship. Nabeel Hamdi put it very well: "The best processes ensure that all concerned will share the responsibilities, profits, and risks of what they will decide to do."[20]

Politicians typically find the concept very risky, and even (as the preceding short section about politics suggests) consider that it is a negation of their duty to put the onus on the electorate to take decisions when they (the electorate) have elected others to do so. However, there is something reassuring about the term *partnership*, which implies a different relationship: the "authority" takes the initiative to enter into a partnership with another body. Whether that body is a community, a private corporation, or another public sector entity, the term *partnership* is finding increasing currency as a developmental concept.

There are two very important differences between the concept of partnership and negotiation. The first is that partners choose each other, and second there is mutual dependence. Exploring these two differences will help to illustrate the nature of the relationship.

> *Choice.* As marriage counselors know, choice does not preclude conflict, but the fact that a relationship starts out of choice creates a dynamic that differs from one where the parties are forced to, for example, negotiate in order to avoid conflict. This is not to say that the terms of partnerships will not require negotiation, nor that conflicts will not arise within partnerships. However, the relationship in a partnership starts from a position of basic trust, as without that trust the choice would not have been made.

> *Mutual dependence.* Mutual dependence is not a financial relationship but a tactical one in which the parties decide that they can achieve more if they are together than apart. But it is often much more than that: the parties recognize that they are different and bring complementary attributes to a relationship. Mutual dependence can exist without the parties having equal power, but it flourishes where there is symbiosis.

This mutuality may have long roots, or it may develop from specific circumstances. An example of long-term interdependence is the relationship between trade unions and employers. In our well-worn example of the railway company, one could never claim that the parties had a mutual dependence from the start. In that case the mutual relationship arose from a need to solve a problem which affected both parties—and therefore an interdependence. Whatever the origins of the mutual dependence, recognition that it exists is the first stage in building a partnership. Partnership is a working relationship; interdependence is a condition.

What has become clear is that where mutual interdependence exists, helping to move it into an active partnership is the first stage of fruitful and real participation. The remarkable feature of a partnership is that it removes the sense of patronage that usually clouds so-called participatory projects. If the concept of partnership can be used to forge a relationship which has a business rather than a social purpose, then the preconditions for success can be established. The merit of the business approach is that it is relatively value free—by contrast with the social model, which implies that there are those who need help. Help implies patronage, and patronage implies that the patron knows best.

Let us pause here to examine this in more detail. At first sight the social approach is exactly what is required. It addresses the need for clean water, good roads, health services, and so on. How can one dismiss these needs? Is this not a negation of what we, as development practitioners, have been fighting for—for so long? Is this a rejection of the Millennium Development Goals? Are we saying that the struggle for improved services, in which so many residents have been engaged, should not be the starting point for development?

As much as these rhetorical questions are designed to make the concept of abandoning social goals, or ignoring the need to help the poor, look ridiculous, the point here is to look at means, not ends. One can never question the need to facilitate (this is deliberately selected as a neutral term) the improvement in the lives of the poor. But how those ends are achieved is what we are looking at.

For the purpose of argument, the social approach can be characterized as the top-down method. We know that conditions are bad, and we know that people want improvements. So let's just get on with it. In the process we should consult the beneficiaries in the knowledge that what we are proposing is what they need. What is wrong with that?

Two things. The first, which we discussed in Chapter 4, is the deterministic model of project design and implementation. There is no need to rehash those arguments here. The second is that this starting point sets up the wrong relationship, which is one in which there is a donor and a recipient, and this relationship is fundamentally one-sided.

Business relationships are different. The fundamental of business is that one party has a commodity for sale, and another wishes to buy it. Whether the commodity is ice cream, high-powered consulting services, or my secondhand car, the relationship is built around the concept of freely entering into a contract.[21] Buyer and seller have exactly the same power in that contract—that of choice as to whether to enter into it or not.

Thus if we characterize the relationship as a partnership, we are looking at a mode of working which implicitly and explicitly recognizes the equality of the partners, just as if they are under a contract.

There is no magic wand in partnerships, but the concept brings with it power for both parties, and assists everyone to perceive the relationship as a businesslike one.

COMMUNICATION FOR EFFECTIVE PARTICIPATION

There is another aspect of participation which is greatly undervalued: communication.[22] For many, the need to communicate information is seen as an obvious one, but in practice there are many problems.

The need to have a strong knowledge base for informed decisions to be taken is a matter of concern. Central to all experiences of effective participation is the concept of communication. In South Africa, the Mvula Trust evaluated 68 community water supply and sanitation projects in terms of their sustainability. One of the most important findings was that good communication and information were essential.

> Innovative formats for information dissemination were discussed in some detail in the evaluations. A number of options were put forward, including the idea of "visual accounts" for financial information. They said the design should be simple and colorful . . . possibly drawings could be used to assist illiterate members of the community. The visual accounts should be posted prominently on community notice boards or taken from the household to household . . . It was agreed that more creative ways could be found to disseminate general project information. Suggestions included billboards, posters and community radio . . .
>
> One example of innovative communication at project level comes from Operation Hunger (also a South African NGO). Billboards were used in Klipfontein in the Northern Cape to increase community awareness of nutrition problems in the village, and to monitor progress in addressing nutritional problems as they emerged. The result was a well-informed community who reacted to emergent problems, and a reduction in growth faltering over time. The information was color-coded and designed by a local (illiterate) resident. Evaluations demonstrated that the billboards increased local awareness considerably . . .
>
> In short a few clear lessons have emerged about communication at project level. The first is that Mvula's assumption that working with Village Water Committees as a centralized community representative structure that will disseminate the necessary information and obtain community input to decision making is, quite simply, wrong. Where communication and decision making has been decentralized, it has been far more effective. Project agents, both social and technical, will

have to work directly with community members to ensure proper input to decision making and information dissemination.[23]

The impact of poor communication is illustrated in the following example of an upgrading project in Nairobi which was initiated by a development agency through the local church.[24]

A trust (Amani Trust) was formed to serve as the developer and the implementation agency. They received the title to the land from the City Council, and then bought the structures from the property owners for an agreed amount of compensation. Once the property belonged to the Trust, they could conduct the upgrading in a phased manner. In principle, the components of this project follow the basic fundamental principles on standards and so on discussed earlier: roads, water supply, drainage, sanitation, street lighting were all improved, and all concerned parties were in agreement on those aspects when the process began. However, somewhere along the way, a misunderstanding developed about the property tenure: the community says they believed this was a mortgage program and that "the houses would be theirs after a certain number of years." According to the Trust, this is—and was intended from the beginning as—a rent-to-own scheme, and alleges that the misunderstanding was caused by false promises made by some politicians. Whatever the reason, this combination of political instigation and inadequate communication between the two parties led to a situation of violent protest and vandalism, court rulings, and police interventions in 2000. The project has since been stalled.

Let us now return to the key issue of the ingredients of good project support communication. There are several aspects to the question:

- What should be communicated?
- By whom?
- How?

What?

Authorities typically like to withhold information that they think could be divisive or politically charged, such as information about charges, rates, and taxes. They fear that if they engage the population on such matters, they will generate protest. They are very reluctant to subject their way of doing things (e.g., standards of road widths and construction) to public scrutiny, and will not engage on subjects such as minimum plot size, and so on. It may not be fair to call this secrecy, but this is how it is perceived. As a result, suspicions are aroused, and even if the authorities have no or little room to maneuver, the people whom they are serving have a right to know what is being proposed and why. This is therefore a fundamental point.

The authorities very often need to understand how they are perceived. In Zambia we had staff with video cameras who interviewed community members regarding the implementation of the project. Criticisms of the staff turning up late to work, mismanaging allocations, being rude to the public, and so on were very effective in getting the attention of the management as well as the staff concerned. Other concerns being felt by the public also have to be communicated, and while much of the participatory process can remain at the field level, much also needs to be shared at the civic level.

Communication must therefore be a two-way process, no matter how uncomfortable some of it may be.

By Whom?

If there is a need to communicate this information, who should do it? There is a great deal of value in having a neutral person look at the communications needs from the outside, that is, neither a member of the community nor linked directly to the authorities. This allows the person to be seen as an honest intermediary who is not trying to advocate any specific agenda.

How?

Each message has its own appropriate medium. The system so beloved of politicians—the mass meeting—is just about the worst medium. Difficulties in hearing the message, in retaining the information, and the lack of opportunities to clarify any points of doubt make it an inefficient tool. However, it has its value, and meetings in general provide an opportunity for interaction around the issues raised. Our caveat is that large meetings are much less satisfactory than small ones, and small ones which have the specific opportunity for face-to-face engagement around the issues are the best.

But meetings lack the certainty of print, and printed material is of great value in providing information in an incontrovertible way. Unfortunately, this very quality is what sometimes makes it difficult to produce, as stakeholders argue over precise terms and expressions, or are reluctant to make commitments to specific actions or details.

The mass media have a very important role to play in generating support for a project and informing people regarding it. The difficulty is often that journalists, like many qualified people, have negative attitudes to informal settlements. Attempts to involve them may backfire if not properly handled, leading to headlines such as "Eyesores to remain—government abandons plans to remove slums." For this reason journalists need to be given an insight into what is being done and how it benefits the whole community.

Television and radio are, of course, superb media for human interest stories. They can also be used to debate issues, so that senior local government representatives can be exposed live to the concerns of the poor.

PITFALLS OF PARTICIPATION

Without leadership, community action will flounder, but there is good and bad leadership. Bad leadership can destroy vertical and horizontal relationships and neutralize all efforts to build a genuinely participatory relationship.

The three most common forms of bad leadership are abuse of power, corruption, and crime.

Abuse of Power

One of the strange features of nature is that plants reflect the conditions in which they grow. In Europe, stinging nettles will flourish in damp and dark corners; in the tropics, similar conditions may foster aggressive creepers which smother other plants. To continue this metaphor, is it possible to cultivate the social soil or manipulate the human environment so that the nettles and creepers do not flourish? Equally, is it possible to put in place conditions in which social pathogens can flourish?

An important insight into human behavior has been given to us by the Stanford experiment in which 24 psychology students were invited to engage in an experiment on the psychology of imprisonment. About one hundred students had replied to an advertisement offering a small payment for participating. Those with criminal records or medical or mental problems were screened out. Those selected were invited to choose whether to play the role of guard or prisoner, but no one wanted to be a guard. A random process was used to select the guards.

At the start of the experiment, the guards were invited to choose their own uniforms, and participated in the last day of building a mock prison in the basement of the psychology faculty. They were then briefed that their duty was to maintain law and order, not to allow any escapes, but they were not to use violence against the prisoners.

The prisoners were told to wait at home, and make themselves available to participate in the experiment on the first day—a Sunday. In fact, the study organized arrests of the "prisoners" by real policemen, under the glare of television cameras, following which they were driven first to a police station, where the full ritual of fingerprinting and so on was followed. Then they were taken to the fake prison and put behind bars.

The guards were given reflective dark glasses to enhance their authoritarian image, and given billy clubs, handcuffs, and whistles. By contrast, the prisoners were dressed only in smocks, with their hair covered in women's stockings, thus reducing individual identities. A chain was attached to their ankles.

From the start the guards devised ways of stamping their authority on the prisoners—making them say their prison number fast and clearly, reciting prison rules, and so on, and making them do press-ups if they were seen to grin or in other ways seem to defy authority, and so on. Soon, the guards seem to compete with each other to be stricter and more authoritarian.

They include disturbances to sleep as one of their tools—waking up all the prisoners when the guard shift changed at 2.00 a.m. and making them perform meaningless counting and press-up rituals for an hour. By the morning of the next day it is clear that the guards are enjoying their power.

Confinement in an unlit room—called "the hole"—is used as a weapon against rebellious prisoners. After all prisoners have been told to make the bed, one has the sheets and blankets pulled off and thrown on the floor—a fist fight ensues and the outraged prisoner is taken to the hole. Later that morning there is more rebellion as the prisoners fight against the arbitrary and meaningless rituals forced on them by the guards. The guards respond by dragging the blankets through burrs—forcing the prisoners to spend hours picking them off. As a sign of rebellion some prisoners tear the numbers off their smocks—they are then forced to go around naked, as a punishment. Beds are removed in another punishment. Some prisoners barricade themselves into their cells, as an act of defiance.

Within 36 hours one of the prisoners was in such obvious mental distress that he is released. He had been interned in the hole many times, and had been disciplined in numerous ways.

By Thursday evening there are only five prisoners left—the others have had to be released before the experiment was completed. The situation gets worse. Prisoners in the hole go on a hunger strike, which enrages the guards. The guards humiliate them with new and more extreme games, such as one where they force the prisoners to simulate animals in sexual activity. That same evening a fellow member of the faculty witnesses the experiment and is totally horrified at the abuse that she sees. She persuades Philip Zimbardo, who is conducting the experiment, to abandon it—less than halfway through its intended duration.

Zimbardo writes as follows:

It is hard to imagine that such sexual humiliation could happen in only five days, when the young men all know that this is a simulated prison experiment. Moreover, initially they all recognized that the "others" were also college students like themselves. Given that they were all randomly assigned to play these contrasting roles, there were no inherent differences between the two categories. They all began the experience as seemingly good people. Those who were guards knew that but for the random flip of a coin they could have been wearing the prisoner smocks and been controlled by those they were now abusing. They also knew that the prisoners had done nothing criminally wrong to deserve their lowly status. Yet some guards have transformed into perpetrators of evil, and other guards have become passive contributors to the evil through their inaction. Still other normal, healthy young men as prisoners have broken down under the situational pressures, while the remaining surviving prisoners have become zombie-like followers.[25]

Zimbardo has since has written about the experiment at length, and drawn comparisons between it and many other situations. From his work, and that of many other psychologists, it is clear that it is the freedom to use power that acts as a drug liberating something abusive within us. As soon as the constraints of normal society are removed, we all have the potential to commit horrific wrongs. The Rwanda genocide, the guards in Hitler's concentration camps, the Pol Pot purges, the guards at Abu Graib, tyrants in many different ages and from many different countries: these are all evidence of the same phenomenon.

These comments may seem somewhat extreme in the context of a book about development, but they are pointers to the risks of leadership which is not constrained by the checks and balances of good governance and social values. As soon as a leader senses that he can do what he wants, then abuse may start.

Similar abuses of power were made possible in the 1970s when the South African government started to implement its plan to create tribal homelands for all black people in South Africa. The white government cared little about how those homelands were governed as long as they created no trouble, and could be presented to the world as the choice of the people concerned. The leaders of the homelands were given generous funds with which to build grand headquarters and maintain law and order. Normal checks and balances were not applied—that was considered "their" problem. As a result they were, almost without exception, tyrannies.

Crime

In many communities the old-fashioned crimes such as theft and murder are being superseded by the modern crimes around drugs and prostitution with their concomitant gang warfare. Due to the illegal but very profitable nature of these activities, it makes economic sense for the operators of these services to create monopolies which must be protected by force. This use of force to protect economic interests is often the most damaging manifestation of crime. The growth of the mafia during the U.S. prohibition was one of the first examples: these days it is usually drug cartels. Once violence becomes a feature of a neighborhood, then protection rackets can be introduced.

Janice Perlman undertook a study of 750 residents of the *favelas* of Rio de Janeiro in 1968–69. Thirty years later she returned to study the same people and find out what had changed in the environment in which they lived as well as their lives and that of their families.

This presentation is based on preliminary findings from a restudy of the people and communities described in her 1976 book *The Myth of Marginality: Urban Poverty and Politics in Rio de Janeiro*. The original research involved living in three communities and interviewing 200 randomly selected residents and fifty leaders from each. The first community, Catacumba, in the upscale residential South Zone, was forcibly removed in 1970 and the residents relocated in public housing projects (*conjuntos*) around the city. The second, Nova Brasilia, in the industrial North Zone, is part of the now notorious Complexo de Alemao, which is one of the last areas untouched by the widespread upgrading project, Favela-Bairro. In the third site, Duque de Caxias, in the peripheral Fluminense lowlands, half of the interviewees were *favelados* and half owners of small unserviced lots in the poorest areas of the municipality.

One of the most striking changes in the 30 years that had elapsed was that the community cohesion and mutual support that had characterized the *favelas* of the 1960s had been replaced by fear. In the 1960s there had been fear—that of being forcibly relocated by the authorities to a remote site; that fear bound the communities together in a sense of common interest. This had been replaced by a very different fear—that of being caught in the cross fire between police and gangsters or between rival gangs. Perlman says:

> The people we interviewed were afraid of dying every time they step out of their front doors, and they fear their children will not come home from school alive. Even inside their homes they do not feel safe. At any moment they fear the police may kick in their doors on the pretense—or reality—of tracking down a dealer. Alternatively they fear someone fleeing the police might put a gun to their heads and insist on being hidden, fed and housed . . . In the 1960s a few outlaws in Rio had handguns, but now all dealers carry automatic weapons . . .
>
> The new reality is reflected in the lives of the *favelados* in multiple and pernicious ways. Most importantly, the very communities in which they were trying to lead normal lives have become "contested spaces." Increasingly occupied by mid-level dealers and their legions . . .
>
> The pervasive presence of the dealers has had devastating effects on community life. Compared with thirty years ago there is considerably less "hanging out" in public space, less community participation in community associations, and (especially where there is a war between *commandos*) less visiting among friends and relatives. Membership in every kind of organization, with the exception of evangelical churches, has declined drastically. The internal space of the community is no longer the locus for leisure or recreation. These were the things that formerly united and bound the community together.[26]

If criminals stuck to crime, the problem might be less severe than it is. Unfortunately they need to protect themselves from scrutiny, and to allow themselves the most freedom to operate without scrutiny or interference. In order to achieve this they must also interfere with regulation of all sorts and policing. Transparency is their biggest enemy. They operate as insiders—and only the insiders are privy to what decisions are taken and why.

Figure 6.3   Scenario of area dominated by drug gang.

The Railway Company	The Community (dominated by a criminal gang)
Railway company makes approach to nominal leader of community. Community decision has to be taken: do we talk to the railway company?	Decision referred to inner circle of gang.
Inner circle of gang summons leader to ask what the issues are: what is the railway company wanting?	They are told about the redevelopment scheme, which will resettle all residents.
Discussions reveal that they will be removed from inner city.	Boss decides the move will destroy their market and fears that resettlement will allow authorities to screen applicants and could reveal true identities, etc.
Community leader told to inform railway company that they refuse.	

If a community decision is being made in an area dominated by a drug gang, for example, it does not require much imagination to construct a scenario like that shown in Figure 6.3.

Any party wishing to engage the community therefore faces two very substantial difficulties. The first is that he or she cannot discuss the issues face to face with the boss who, for many very good, practical reasons, does not want to be either identified with the decision, or even identified at all.

The second is that, in any case, the interests of the community are subordinated to the interests of the gang: therefore it is pointless to even talk to the community because they have no say.

The participation rule book therefore has to be thrown away where communities are dominated by gangs. Communities dominated by gangs may be reluctant to admit it, but it is usually quite well known where this is happening.

In such cases, different tactics must be employed. These are not the tactics of the well-meaning civil servant or development NGO, but that of a cunning operator. He or she must first get close to the boss and then start to negotiate. There are examples of gang leaders and their members being brought into negotiations as hostile and suspicious parties who, after sustained and effective negotiations, come to support genuine change which, at first sight, is against their interests. What the negotiator has done is to help them consider alternative scenarios in which they can see a bright future. There is no recipe for this but a smart negotiator will be able to bring in an appeal to civic pride and duty, and the appeal of legitimacy as a preferable lifestyle to crime. But this is not easy, and there is no recipe for success. Perlman[27] writes:

> We found little evidence that drug dealers had set up a "parallel state" of paternalistic benefits for the poor. There is a lot of talk about the new *caciquismo*, wherein the drug lords provide schooling, health care, food, and protection to residents in exchange for loyalty. But this was not the case in the communities we studied. While it is undoubtedly true that some people come to dealers in cases of emergency—needing a ride to hospital for an ailing relative, money for food if they are hungry, or perhaps access to a place in the local school—this is the exception rather than the rule. Only 10 percent said that the drug dealers had even helped them in any way . . . The majority of respondents were afraid to even answer the question.

Sen makes an interesting point regarding criminal gangs as *law enforcers*.

> When, however, the standards of market ethics are not yet established, and feelings of business trust are not well developed, contracts may be hard to sustain. In such circumstances, an outside organization can deal with the breach and provide a socially valued service in the form of strong-arm enforcement. An organization like the Mafia can play a functional role here and can receive appreciation in precapitalist economies being drawn into capitalistic transactions. Depending on the nature of the interrelations, enforcement of this type may end up being useful for different parties, many of which have no interest in corruption or crime.[28]

Corruption

Corruption goes hand-in-hand with crime, but the two have different impacts and outcomes. In some ways crime is easier to address because it is more cut and dried. Corruption can erode more than a few gang members: it can undermine a whole society. It has four very damaging effects.

- In a corrupt society the motivation for a decision is monetary reward, not the merits of the case.
- In a corrupt society, no one trusts anyone else. This means that all decisions are sources of conflict as the parties try to distinguish the motivations of each party and the likely hidden reward that they could get.
- Because a percentage of all funds for development must be used to corrupt those in a position to demand corrupt payments, less money is available for development.

- Because all transactions have within them the potential for a bribe, it is in the interest of the parties to make the transactions seem slow or difficult—unless, of course, a bribe is paid.

Where corruption is truly embedded in a society, then it becomes more than the occasional bribe—it is part of the universally accepted means of accumulating wealth in which not just the corrupted official benefits but his whole extended family. If he fails to earn the revenue that is expected of a person in his position, he will be sneered at as being too weak or ineffective.

The barriers placed in the way of ordinary people conducting legitimate business by corruption are graphically illustrated in this extract about Kibera (Nairobi, Kenya):

> I interviewed scores of people—landlords, tenants, even academics—who all told me the same story: anything that you want to do in the community, from constructing a dwelling to making a serious repair, the civil servants from the district administration (known as district officers, chiefs, assistant chiefs, and village elders) insist on payoffs. If you don't pay them off, they will knock down anything that you have built or call in the police to inspect your business. The going rate for someone who wants to build a new mud hut is 2,000–3,000 shillings: equal to the average Kibera worker's monthly pay.
>
> One resident who was able to buy his hut lives with a plastic tarp strung underneath this rusted-out metal roof, because he said it would take too much in bribes to fix. "To repair the roof, first I would have to see the chief," he told me. "It would cost 2,000 shillings there. Then the village elders. Maybe 1,000 to be distributed among them. So I'm better off leaving things the way they are."
>
> Another person, a Kibera landlord . . . said bribery was a way of life. "If you want to survive you have to cooperate with the provincial administration and the police department," he told me. "Let's say if you profit 100,000, almost 40,000 goes to them. If you refuse, your business is history."[29]

Box 6.6 and 6.7 presents a similar situation in Kenya and Nigeria. Our case studies in Chapter 3 provide further examples of the regular harassment that residents face, due to the so-called illegality of their development. For example, Rose's community routinely raises a levy to pay off the municipal officials who threaten to demolish their houses.

Motivation in a corrupt system is treated with a huge sense of cynicism. For example, founding a church or a NGO is seen—with justification—as an excellent step in getting rich. Box 2.9 in Chapter 2 illustrated

Box 6.6   Corruption in the system: A case in Nairobi.

Here we continue Desta's story from Box 4.7. Desta came to Nairobi with KSh140 and today successfully manages and owns 4–5 separate businesses. Through hard work and diligence, Desta has tried to make life better for herself and her children. But she has not has it easy. "Every step of the way, there's someone to stop you, someone to pay bribes to."

"In case of a problem," she says, "we go to the Chief. But if you don't produce something (bribes), there is no help in that office. The Chief sends me to the Chairman, and the Chairman sends me to the Chief. Shuffling between the two, we can never get anything resolved, so we are forced to help ourselves." Desta says she has never seen the NGOs or benefited from any of their programs. "When they come, they are taken by the 'leaders' to those who are friendly to them. Corruption has never finished! Food aid comes, but there is discrimination in distributing the food. People who really need it don't get it. Our leaders don't stand the right way; they stand with those who can give them money."

"The Chairman does not get paid an official salary . . . so where does he get all his money from? If you want to rebuild your house, first, you must pay money for approvals. Second, you must hire the workers assigned by the Chairman to do the construction; you can't build yourself. When you go to ask for land, you pay. There are no receipts for anything. And no record that this land belongs to me. Unfortunately, our leaders are the biggest problem. There's a selling price on everything, even donor assistance, as determined by the corrupt leaders."

"We have no MP here, no representation in government. The Chairman is elected, supposedly. But we are not always called to vote. The last election was in 2001, which was only for name's sake. The Chief has the ability to overrule the election results, and appoint someone. You get elected if you bring money to the Chief. We can't protest because we will get beaten, our houses will be burned, or something else. We fear. . . . !"

"The youth cannot do much with this type of leadership. They suffer from the same bureaucracy: Chief to Chairman, Chairman to Chief. It's a case of the blind leading those who have eyes in this case. They are illiterate. And if one is a bit literate, they ensure you are kept down, and what you want doesn't get done." With regard to the health situation in the settlement, she says, "Maybe as many as 40 percent of the households have someone seriously sick, possibly with AIDS. But the medicine that comes, does not go to those who are really sick."

"There is an organization where you can go to get AIDS drugs, but most people don't know about it, what it costs. Besides, before you can go there, you need to get a letter of approval from the Chief. So who will go, especially when you want to keep these things confidential?"

Source: Ashna Mathema, *Nairobi's Informal Settlements: A View from the Inside*, Background Paper [Unpublished background research funded by the Norwegian Trust Fund/The World Bank], 2005.

one aspect of the collapse of trust in the real estate sector in Nigeria. In a similar vein, Smith writes:

> Widespread cynicism about the ulterior motives of even seemingly well-intentioned endeavors permeates popular culture, with almost anyone or anything being a potential target for an accusation of 419[30] ... Churches and NGOs are supposed to be institutions in the service of the greatest good. The fact that they are often perceived to be frauds attests to the depths to which Nigerians believe they have fallen. In recent visits to Nigeria I have heard civil rights leaders, anticorruption crusaders, and even outspoken individuals living with HIV as being perpetrators of 419, because they are suspected of manipulating appearances purely for their own gain.[31]

Smith (2006) says, also on Nigeria:

> To the extent that Nigerians are participants in corruption as well as critics and victims, it is because they are pragmatic: the stakes for individuals in Nigeria are tied ideologically and materially to the social

Box 6.7   Corruption and crime: A lethal combination.

Nigeria gives us an example of an even worse mixture: crime and corruption. The corruption in Nigeria – widespread as it is – has been a source of huge concern to many people who perceive that its slow pace of development, in spite of huge oil wealth, is attributable primarily to corruption. When a vigilante group called the Bakassi Boys was formed, therefore, to fight crime and corruption in the police, the news was greeted with great excitement. They were a gang which dressed in black, wore dark glasses, and carried powerful weapons.

Initially they were so effective in reducing crime that they began to get state support. They emerged as folk heroes, and were featured in movies and comic strips. The public began to attribute supernatural powers to them, in terms of their inability to be harmed by bullets, and to distinguish the guilty from the innocent by simply looking at people. They conducted popular mass executions of criminals they had arrested, and typically shot to kill anyone they thought was a criminal.

However, their power soon attracted the politicians who saw an opportunity to use the Bakassi Boys to eliminate opposition or threats to their status. The Bakassi Boys also started abducting people and demanding ransoms. They demanded payments to sustain their operations from businesses, and any businessman not paying his contribution was ruthlessly beaten up. In the end the popular support which had prevented the Federal Government from clamping down on them eroded as news of the atrocities leaked out, and their headquarters were destroyed and many of them were killed by the police.

Source: Daniel Jordan Smith, 177–188.

groups to which they belong. Thus when individuals make choices that one might describe in terms of corruption, they do so with a sense that their own failures to acquire resources will drag others down, and with the knowledge that their own success will be evaluated in terms of its contribution to the larger group. Further, people are well aware of the intense scrutiny they face from their families, communities and other associates. Hence, when Nigerians refer to the Nigerian factor, they are referring not only to corruption per se but to the pragmatic choices that individual must make in the context of their obligation . . . to deliver to their people whatever share of the national cake they can capture.

Figure 6.4 is an illustration of the dynamics of decision-making in a corrupt system.

Figure 6.4 Dynamics of decision making in a corrupt system.

Railway Authority	**Local community with corrupt leader**
Railway company want to move residents from land, approaches local leader.	Local leader insists that he can take decisions on behalf of all residents.
Railway explains plan to resettle all residents.	Leader says it will be a tough decision to sell; he will have to employ a lot of people to explain it.
Railway pays him the money knowing that most of it will go into his pocket.	Leader employs young people to inform everybody that they are to be resettled, and that he is negotiating for the best terms.
Railway receives petitions from angry residents who suspect that the local leader has been bribed.	Local leader says he can deal with the troublemakers, but that he will need cash.
Railway pays him more money.	Leader gives cash to troublemakers: 50% down now, and the rest when resettlement has happened provided that everyone moves peacefully.
Railway gets call from other leaders claiming that they are the real leaders. They demand special big plots in the new site and other privileges. They also say they have large expenses to pay, and threaten that unless these payments are made they will ensure that there is no cooperation.	Additional leaders are paid out. Other community members find out about the payments from accounts clerks in the railway, and go on protest march.

This simple example illustrates how the lack of principles on which decisions are taken undermines the spirit and purpose of any participatory process. In using financial reward as the only real criterion for a decision, the leader has lost his power. Corrupt individuals are very vulnerable for this very reason, which makes dealing with them even more difficult.

Corrupt officials or community leaders fear transparency more than anything else, so the starting point must be to disseminate information as widely as possible. The second is to involve as many people as possible in decision making. While this seems cumbersome, it pays handsome dividends in the long run.

However, at this point we are less concerned with operationalizing participation than understanding how it works and what it represents. Operational issues are covered in the next chapter.

Delays and Deadlines

One of the criticisms of participation is that it takes so long. A popular trick in negotiations is to impose a deadline, because without deadlines hardliners like to drag their feet.

There is much merit in this principle, but there are also dangers. If the deadline is too short the whole process becomes a sham. People must have the opportunity—which means time—to reflect on the options they have. A decision taken in haste may be regretted at leisure, as participants in shotgun marriages often discover.

This does not mean that the process cannot be businesslike. In a well-managed environment it is possible to establish norms for how long decisions take, and build project implementation around such norms. In this way, for example, a period of three weeks can be provided for neighbors to agree on the route for a new access path. If they genuinely cannot agree within that period, then mediation might be needed, but in general it is useful for them to know how long they have and to work around that.

But for many politicians participatory procedures seem like an indulgence. "Don't worry," the leader will say, "I can get a decision in no time." A deadline will then be set around that claim, and in the hurry people's opportunity to discuss issues and reflect on their decisions is lost. The short-term gains as a result of quick decision making are usually far outweighed by long-term delays and costs as a result of inadequate participation.

PARTICIPATION: SUSTAINABILITY, SELF-RELIANCE, AND SELF-RESPECT

After outlining the risks to participation, it is only right to return to the rewards.

There is an aspect to participation which is very important. Giving responsibility to the residents enhances their self-respect. To return to

Amartya Sen's advocacy for development as freedom, he demonstrates that economic development gives people the freedom to make choices, and freedom from the millstones of malnutrition, ill health, and the drudgery of daily life which the poor have to face. To take an obvious example: the poor may either have to carry water over long distances, which can eat hours out of every day, or must pay for it from water vendors who charge 40 times as much per liter as the water company.

The involvement of the community in the solution of problems such as this gives them a self-respect which enhances their feelings of self-worth, and demonstrates that even if they are poor they have the freedom to choose which solution to adopt, and how much they want to pay for it.

To end, it is appropriate to give an example which may demonstrate better than any other the true meaning and value of bottom-up development. It is ironic that while we professional people write as if development is a technical process, it takes a journalist to bring out the fact that the human dimension is far more important.

The story concerns sanitation. Villagers in Bangladesh have suffered for years from the diseases caused by poor sanitation: indeed, 24 percent of the deaths in Bangladesh are caused by waterborne diseases. Most people would say that solving that problem is easy: you just build toilets. Ambitious subsidy programs were developed and people were given subsidies to buy new toilets, but many chose not to do so. Also, the new design relied on water, which was not always available.

The problem was therefore how to persuade people to change their habits, and, not only that, how to persuade people to spend money on improved sanitation. And by habits was meant not just their toilet habits, but washing hands and similar matters of personal hygiene.

WaterAid supported the Village Education Resource Center (VERC) to follow a very different approach, which was to help the community understand how their behavior was polluting the water, and the tons of fecal matter that they were depositing every year around their water sources. This was not done by sophisticated multimedia presentations but by helping the individual family think about the situation and understand the risks that their present behavior was causing. Although new sanitary systems would cost money, so did sickness—some households were spending 20 percent of their income on antibiotics and oral rehydration therapy every month. Productivity fell when farmers were too ill to work in the fields. Education suffered as children were too ill to attend school.

The story told by Barney Jopson[32] is that of one of the villages which have been supported by VERC in the 100 percent sanitation program. In this case one member of the community took personal responsibility to convert all the residents into wanting and creating a 100 percent sanitary environment. He persuaded people to dig their own pit latrines, and brought the children into the task of acting as monitors, reporting anyone who defecated outdoors. His tool was simple: the evangelism that persuades people to change their minds. But it was evangelism enriched by plenty of evidence, and he achieved 100 percent success.

What is more remarkable is that this program is being widely replicated —a total of 2,414 communities have now been declared free of open defecation. As more communities become convinced and see the benefits and copy it, so the message spreads faster and more effectively.

The journalistic account skims over the methodology which has been developed for the program, which is basically that of helping community members to understand the facts and work out their own solutions.[33] It sounds simple, but requires a special sensitivity and a methodical approach which encapsulates much of what this chapter has been advocating about the development *process*.

7 Crossing the Great Divide
Negotiation and Consensus Building

This chapter is about the need to adapt our way of working to the needs of the clients we serve. Moreover, in working in informal settlements we are intervening in an environment which has been developed by the community concerned, and over which they currently exert control. Whereas the previous chapter looked at principles and concepts around participation, this one looks at implementation issues. It describes ways of making participation work, and of dealing with difficult situations which typically occur. Managing participatory projects requires special skills, and the chapter concludes by describing the sort of professional skills required and describes training techniques which can help to generate such skills.

There have been plentiful references in earlier chapters about the need to change the ways things are done. We have suggested that participation by the community is essential, and that the public sector should adapt its way of working and its systems to the needs of its customers as a whole. Typically, government procedures are best suited to the well-off; those who are literate, have bank accounts or staff to help them, and are experienced in the ways of the world. These same procedures are, normally, worst suited to the needs of the poor.

There are interesting parallels with business here. For a long time business looked at the poor as a residual component of the market as a whole, to whom little attention should be paid because they consume so little. Of course this is completely true when we consider the market for new cars, but is it so true when we look at the market for detergents, matches, shoes and clothing, and so on? As individuals they buy less, but as a class they buy more. Shouldn't there, therefore, be an approach to the market of the poor which looks at their needs? This is what an ex-Unilever man, C. K. Pralahad, noted when he talked about the "fortune at the bottom of the pyramid."[1] For example, in the field of daily consumption needs, the packaging is often such that the very poor cannot afford to buy so much at once. Breaking down the packing into much smaller, and therefore more affordable, units is described in the following passage:

The logic is obvious. The rich use cash to inventory convenience. They can afford, for example, to buy a large bottle of shampoo to avoid multiple trips to the store. The poor have unpredictable income streams. Many subsist on daily wages and have to use cash conservatively. They tend to make purchases only when they have cash and buy only what they need for that day. Single-serve packaging—be it shampoo, ketchup, tea and coffee, or aspirin—is well suited to this population. . . . For example in India, single serve sachets have become the norm for a wide variety of products, as shown in the table below. . . . The format is so popular that even firms producing high-end merchandise have to adapt it to remain viable long-term players in the growing markets.

Single-serve value at retail

Rs	$	Typical products
0.50	0.01	Shampoo, confectionery, matches, tea
1.00	0.02	Shampoo, salt, biscuits, ketchup, fruit drink concentrates
2.00	0.04	Detergent, soap, mouth fresheners, biscuits, jams, spreads, coffee, spices
5.00	0.10	Biscuits, toothpaste, color cosmetics, fragrance, bread, cooking oil, skin cream.

Measured in tons, the size of the Indian shampoo market is as large as the US Market. Large multi-national companies, such as Unilever and Procter and Gamble, are major participants in this market, as are large local firms. Because the poor are just as brand-conscious as the rich, it is possible to buy Pantene, a high-end shampoo from Procter and Gamble, in a single-serve sachet.

This extract is taken from a book which painstakingly illustrates how adaptation of the way that things are done can totally change a market or service.

In the world of finance, the Grameen Bank has created many important precedents. Of particular interest were two aspects: first, the bank respected the women concerned, and, instead of dismissing them as too poor and noncreditworthy (as conventional financial institutions had done), it *trusted* them. Second, it recognized that even though small sums of money were involved, it was possible to serve such a market and make money if the systems were right.

The HFC Bank in Ghana has made a similar breakthrough, not in loans, but in savings. As Pralahad noticed in terms of the retail trade and services, the individual sums are small, but cumulatively they can be large. In this instance, we are referring to the cumulative effect of small savings. In the markets of Accra, Ghana, where there is a small daily turnover, the market women found it very difficult to save. Saving was especially important for some who had to buy stock in bulk from wholesalers; and any attempt to expand was thwarted by a lack of capital. The response of the bank was to send their staff to *collect* the money from anyone who wanted to save, on a daily basis. There was no compulsion regarding the amount or the frequency of payments, but people who joined the scheme all found it much easier to save by giving it to the bank staff every day. The collectors were young people who were selected for their friendly and courteous behavior. They knew all their clients by name, and when the clients had no money to save would stop and discuss the reasons, and counsel them. In this way the bank became part of a social network—widely respected and trusted (see Boxes 7.1 and 7.2).

As a second stage people who had a good savings record went on to borrow. In some cases the sums have been quite large (e.g., $20,000), but there have been no defaults. In spite of the apparently labor-intensive nature of the operation, it is one of the most profitable cost centers of the bank. Interestingly enough, the branch from which the operation is conducted has no frills—indeed, it is simply a modestly equipped shop (with two computers and some basic furniture) right next to the market.

There are lessons to be learned from this. The most important part is that the scheme was devised *in consultation with* the market women. They said that they wanted daily collections, and they said that the collectors should be women because they didn't trust men. The second lesson is that the two parties trust each other completely, and witnessing the relationship between the collectors and the clients demonstrates how the close relationship works for both parties.

Moving from the field of commerce and banking to development, is there a way in which things can be structured to work for the poor?

In Chapter 6, we suggested that there were a few key points which seemed to make a difference, such as good two-way communications and the use of word of mouth and small groups so that people can ask questions. We have also tried to show that genuine, bottom-up decision making results in a product that is more cost-effective and better suited to its purpose than the typical professionally designed schemes.

It would be very naïve, however, to suggest that such matters can happen on their own. To be effective, they require a structural adjustment in the relationship between the parties—typically local government and the community. This relationship affects how projects are designed and funded, how communities become involved, how officials interact with the community, and how grassroots community relationships impinge on the role of elected political representatives.

In this chapter we therefore look at these aspects and see what impact they should have on project management, staff skills, and operational systems. In the chapters that follow, we examine the question of monitoring, evaluation, corruption, and discipline.

Box 7.1   Savings and microlending program for market women in Ghana.

The HFC Bank in Ghana Bank offers a microcredit program for market-people, primarily women. It has a savings program, regular contribution to which earns the customers a credit history at the end of the first year. They can then borrow against this record, based on the total amount of savings they have. The annual interest rate is about 30 percent. The HFC staff working on this scheme go door-to-door, or shop-to-shop from one client to the next (daily or weekly, depending on the neighborhood), to collect deposits and deliver withdrawals, essentially serving as a *mobile bank*. Interviews with market women revealed that many of them had very modest starts, with economic conditions comparable to residents of informal settlements in Accra. But today, they have substantial savings, and have been able to incrementally expand their businesses, and even build homes.

The HFC scheme presents a unique and extremely successful case of microlending, which should be expanded in Ghana and adapted in other countries. What makes it remarkable is not just the design of the program—which is unique but by no means the only one of its kind—but the training and attitude of the staff who implement the program. The small team of about 10 comprises young, motivated women, fresh graduates, who are relatively cheap—earning a salary of $400/month which is less than a teller in a bank, but a good salary for a fresh graduate in Ghana. They work hard and are highly respected in the communities in which they work. Each one is assigned about 100 clients, and given the responsibility to do the marketing of the program as they work. So, it is a low-cost program. The interest rate for these short terms loans is higher, so from the bank's perspective, this is a highly profitable business, especially with a default rate close to zero.

From the clients' perspective, this scheme works well. It is essentially: a *mobile bank*, so they do not have to leave the shop to go to the bank to make a deposit or withdrawal; a *safety box*, so they do not have to fear leaving their daily earnings in the shop at night or carrying them back home; and allows for a personal relationship with polite, well-trained and trustworthy staff able to build rapport and trust. The success of the program is evident from two facts: the first was the high level of satisfaction with the program evident among the clients that were interviewed. Second, it allows people to progress economically after a simple and small loan facility: many of the older clients have been able to expand their businesses over the years, and subsequently taken housing loans to build or improve their homes.

HFC staff, in their uniforms.

An aerial view of the market where the program operates. The photo was taken from the balcony of HFC's field office.

Source: Ashna Mathema, *Qualitative Study: Household Interviews, Accra, Ghana*, 2006 [Unpublished background research for African Union for Housing Finance, funded by Cities Alliance/ The World Bank].

Box 7.2 Climbing the informal sector business ladder.

Beatrice, standing by her stall stacked with provisions

Beatrice started out as a hawker in Kanishi selling tomatoes, walking around with a basket on her head. From her savings over the years, she has expanded her business and now trades in provisions (cans, spices, etc.) on a table-top in Agbogbloshie market. Beatrice opened an account with HFC in 2003. Her first deposit was 5K cedis (55 cents). She now contributes 15-30K cedis ($1.5-3) daily, and has got a balance of 1.4 million cedis ($155).

Nancy, counting her earnings of the day

Nancy started out her business in a stall on the side-walk, with a monthly revenue of some 1-1.2 million ($130). Through the HFC program, she was able to save, and reinvest into her business. She bought this shop in Makola market, and increased her stock over the years. Today, her shop's net worth is about 1 billion cedis ($110,000). Her current balance in the savings account is 45 million cedis ($5000).

Above: Annah
Below: Annah interacting with HFC staff, making her daily savings contribution

Annah sells pots and pans. She started this business several years ago, as a small vendor: she would buy goods from out of town, carry them on her head and sell to other market women. She heard about the HFC savings scheme from a friend and opened a savings account some 8 years ago. Her current daily contributions towards the savings account are between 50K and 100K cedis ($5-10), and her account balance is 12 million cedis ($1300). She took a loan of 5 million cedis ($550) last year to buy additional stock, which she has recently finished repaying. Annah says she likes the HFC scheme because "it has savings and loans combined in one place. Plus we get interest on the savings without any deduction or fee, as in the case of susu." With her savings, she has bought a piece of land, and has been incrementally building a 4-room house with a toilet and bath. She is planning to buy another piece of land to build another house for her children, for which she might take another loan.

Source: Ashna Mathema, *Qualitative Study: Household Interviews, Accra, Ghana,* 2006 [Unpublished. For African Union for Housing Finance, funded by Cities Alliance/World Bank]. The names have been changed.

All management relies on delegation to be effective, and the management of participation is no exception. On the one hand, the community must delegate powers to a committee of some kind; on the other, the public sector, which we shall assume in this context is the local government, must delegate the right to interact with the community. This act of delegation poses problems for both sides: the community representatives could assume that they have been given a mandate to take decisions without consultation, while the local government staff can be accused by politicians of acting without approval. It is therefore essential to establish robust structures on both sides which can withstand the pressures and passions that development can generate.

We start by looking at the question of identifying a community structure which represents all interests. The first question here is who identifies it, and what criteria should be used.

IDENTIFYING COMMUNITY STRUCTURES

It would be very naïve to assume that community structures are self-evident, or will identify themselves spontaneously. To make matters worse, many officials are scared of participation, and typically resort to command language as a way of avoiding the—as they see it—complications of interacting with different community interests. In brief, it needs special people to facilitate participation.

The community participation activators have two major tasks initially. The first is how to decide with whom to interact, remembering that in most communities there are many different social networks, each of which has leaders.

Identification of effective and inclusive leadership is probably the hardest task of all. In some societies it is presumed that political structures are the best and only ones;[2] in others it will be religious structures.[3] Often the biggest task is to persuade such leadership that they should bring other leaders into the system, and thereby ensure inclusivity. An exclusive approach to decision making and gatekeeping by power blocks are often the undoing of a participatory process. It is really essential therefore to spend time to identify an inclusive leadership, and to make sure that the process is seen to be transparent. This sometimes requires considerable skill, as rivalry between individuals can be used to destroy a process.

The second task is how to win over the confidence of the people. One of the biggest issues in this context is to handle negative perceptions, as informal settlers have typically experienced negative handling and have heard negative reports regarding the attitudes and methods of the authorities. They will therefore naturally feel hostile, and will have a very skeptical attitude to any person representing those authorities. This may express itself in a refusal to meet, refusing to listen during meetings, shouting and booing, or expressions of total recalcitrance.

In order to win over the confidence of the people, the participation activators must be seen by the community as being neutral, that is, not on either side. If they are perceived to be there simply to "sell" the approach of the authorities, their work will be rejected. Their job is not to tell people what to do but rather to help them decide for themselves. In brief, the activators must be totally nondirective. They therefore have to have supreme patience as they allow people to develop their own solutions.

In this sort of role, even if they think they know the answers to questions—such as why roads are designed in a specific way or how water charges are calculated—they must rather involve a specialist to answer such questions. In turn, such specialists must be prepared to answer questions that, on the face of it, are very hostile; and in so doing they must also be patient.

Such skills are not, as we suggested earlier, a product of a political past, or of a degree in social work or psychology, however much these might inform and help the person concerned. They are, in fact, specific to that type of work. But we will discuss this more in Chapter 8.

STRUCTURING A RELATIONSHIP

We must start by assuming that the initiative is taken by local government which is searching for a body with which to interact. Commonly there are well-established structures: in the old days of the one-party state in Tanzania and Zambia, for example, the party committees offered an obvious point of contact. But even they were not necessarily representative of the interests of the whole community, and we should beware of the quick fix of an existing body which, although it may be the most powerful within the community, is partisan or in any way is not representative.

The first task is therefore to *ask*. A random survey around the community will give a quick idea of whether there are organizations which can be taken to represent the community. During the same informal survey, the question must be asked regarding the organizations to which the person is affiliated. These are typically religious, women's or youth groups, work-based groups (e.g., informal traders), and savings groups. There may be political ones as well. This will then provide us with a universe from which to invite representatives to attend a workshop in which a representative committee will be established. But before any meeting is called, the next stage is to meet the heads of each of these organizations and canvass their ideas about how to proceed. In brief, the process is participatory from the start, but using the gradualist approach to ensure that the right people participate.

How the next stage proceeds will vary. Typically, all interested parties will get together and will form a committee. The degree of formality of this committee may vary: for some it is enough to keep the whole process almost casual; for others there might need to be a constitution which will specify the procedures for elections, management of funds, and other matters. In some cases, the committee will establish a financing arm which can then be used as a recognized body to which public and private funds can be donated or lent (see Box 7.3). This is where the concept of "good government" comes in. For example, it is very helpful if the community can be given support in these matters, and thereby save everyone much time.

As soon as a representative group has been formed, the members need to meet and develop a working methodology. This should encompass two critical components: how to consult the community as a whole before making major decisions, and how to give feedback to the community in terms of progress. These issues are easily agreed at the beginning of a process, but need constant attention if they are to be implemented in practice. This is covered in more detail later.

At this stage, an important technical point must be addressed: What is the role of the political representative usually designated a councilor? He or she will typically be nervous about the role of the committee, as it will seem to undermine his/her duty to represent the community.

This is an issue which must be addressed with care, in two ways. The first task will be to include the councilor in the early discussions regarding the formation of a committee. If the councilor's ward (or whatever the electoral unit is designated in the country concerned) coincides with the boundaries of the settlement concerned, then he or she may already have a ward committee with similar functions. Ward committees exist, under one name or another, in many jurisdictions, and while they look good on paper they are typically marred by the lack of a specific role and the fact that the councilor either chooses the representatives or plays a big role in selecting them. Ward committees therefore can pose a threat to effective community involvement.

For this reason the councilor has to be convinced that the new committee is a development tool, which will help the development of the area. It will collaborate closely with the ward committee and the councilor will be

Box 7.3   Strengthening community groups.

In the USAID Community and Urban Services Support Project (CUSSP) in South Africa, community groups were assisted to form Trusts into which USAID funds could be deposited. A firm of lawyers was contracted to do the paperwork involved for all 28 communities involved, and the process was quick (about 2 weeks) and easy. Similarly, a draft constitution was available which could be used as a basis for each group to develop their own.

represented on it, and is a tool to accelerate development which will be to the councilor's benefit.

To make sure that the councilor is supportive of the initiative, he or she must be fully involved in the process and must see it as a means of advancing her interests.

SERVING THE PUBLIC

In turn, the local government must establish its own presence in the area. Sending in a community liaison person who chooses when to be there, and whose role is to tell the community what is going to happen, is not the idea.

Before we try to define the role of the public-sector team in a community, we should ask what the public will expect of them.

1. They will require good communication. This is not just a question of putting up a few posters, or addressing public meetings, but communicating information in a language and style that the residents can understand.

   Communication can take many forms: for example, in Zambia the project commissioned a top pop group to praise the virtues of alternative technology (compressed soil blocks) in one of their songs, which, incidentally, was at the top of the pops for six months. Street theatre and radio shows can also be effective. The written word can be important, but the spoken word usually has more impact.

   A very important part of this is that all local government staff who come to the site in any capacity must speak with the same voice and either be able to answer questions or be able to say that they don't know and refer the questioner to the right person.

2. They will need to take informed decisions. In order to do this they will need to have not only the correct information but the opportunity to interrogate it. They will not, and should not, have to be given information on a take-it-or-leave-it basis. Experts must therefore be available to respond to queries, whether they are about land tenure, property taxes, road construction, or bylaws. If they do not understand the need for something, they cannot be expected to support it. Cognitive dissonance will set in, and the engagement of the community can be lost.

3. Services should be provided in a way that meets the needs of the residents in terms of hours of operation, location, and the attitude of the staff. We cannot be prescriptive about how to meet these needs, but examples which have been seen to work have the following characteristics.

- The local government has an office—no matter how small—within the community. It can be in an existing house if need be. At the very least, it offers a location to which people can go to make contact.
- The office is staffed by someone from the community, and paid by the local government. This person is basically a liaison person who understands the needs of the community *and* the way that the local government operates.
- The local government team in the settlement can be very small or quite big, but whatever the size, its main function is to act as a facilitator. What it must not be is the means by which decisions made "in the office" are communicated to the people; or, even worse, an enforcement agency designed, for example, to get the people to comply with the bylaws.
- The local government team must have the support of the head office to get answers to questions raised by the community, and to mobilize resources.

In other words, the local government is placing its resources at the disposal of the community to help them develop solutions that it (the community) wants and which work. Equally important is for the team and the community to be given time to take decisions. It is terribly easy, and very tempting, to get an instant decision. Thus, a councilor, for example, may get support by acclamation for a project to bring a new road into the settlement, but this is not a decision that will get the support of the community if they find that a huge number of houses will be demolished to make it possible. A sustainable decision, to which people feel committed, takes time, during which dissonances have time to be resolved.

DECISION MAKING

Community decision making follows a pattern, and in the world of real development we must understand that it is essential to allow the time for decisions to be made, and that time spent at this stage can save far more time which might be required to resolve conflicts at a later stage.

There are three stages in decision making.

The first is to have the information. We talked above about the need for communication. This is, in a sense, the foundation for a process, one which cannot proceed effectively until those involved—which should be every resident—internalize and assimilate the information. However effective mass communication is, people will usually want to ask questions to clarify matters in their mind. They must therefore have an opportunity to question someone about matters which they are not clear about. In the process, trust will develop between the community and the local authority. The best environment for the transfer of information is a small group of between 15 and 25 persons in which people are less shy than in a large public meeting, and quieter people will feel free to ask questions.

The second stage is for people to look at the possibilities and consider their options. Among the types of issues will be the trade-off between standards and costs; the trade-off between the provision of additional facilities and the loss of dwelling units; and the question of the routing of improved roads and the like. This stage should continue at the small-group level, as many questions will arise and debates may occur. The technical nature of some questions will mean that field staff will not be equipped to deal with them; and even if they know the answers, they should consciously decide not to answer them because they lack the authority to do so. Thus the second stage may include some strong questioning and debate around technical issues.

The third stage is the development of consensus. This requires good management so that people whose views are being considered do not feel excluded and resentful. It might need some out-of-the-box thinking. For example, a substantial number of people might prefer to live with only minor improvements to their infrastructure, because they cannot afford anything better. Because the saving in engineering terms is only effective if a geographical area is serviced to a specific standard, those who opt for such reduced standards could consider swapping houses[4] with those who want to benefit from substantially increased standards. In this way, some areas will be serviced to a high standard, and others to a lower one, and charges that each household pays will reflect that.

Consensus building has to be achieved steadily, and cannot be rushed. But we must recognize that there will be some people who will refuse to collaborate, for economic, social, or political reasons. How such people are brought into the consensus model requires tact and persuasion. The objective should be to give the person a voice and then bring social pressure to bear. Often people who are individualistic and stubborn will be willing to sacrifice their personal welfare for the common good if they are given the opportunity to do so voluntarily. But when they are put into a corner, and forced to collaborate, they may make life very difficult (see Box 7.4). Obviously there is no standard method for forcing compliance, but if we refer back to the role of cognitive dissonance, we recall that there must be either a penalty or reward. In this case the preferable route is to reward the person, for example, by huge public acclaim, offers of additional land (or whatever) in compensation, or assistance in kind.[5]

The lessons of negotiation techniques are relevant here. The first point is to recognize that both parties have a legitimate position which must

Box 7.4   Building consensus: A focus group meeting in Swaziland.

I was involved in a project preparation study for the local government of Mbabane aimed at upgrading the informal settlements in the city. We started out by meeting the respective settlement leaders and requesting them to organize focus group discussions to understand the community's perspectives on this initiative. The first focus group revealed complexities, and an indication of what was to come.

The leader got together a group of about 30 community members. We sat on a rock on top of a hill, and I began by describing the objective of my visit. The people asked polite questions, and the responses were well accepted, until suddenly, one person interrupted. He stood up, and started yelling something in Siswati, pointing at me accusingly. My Swazi colleague slowly translated what he said: it was that I was "from the government", plotting to "throw them out of their houses," and that I had come to "divide the community." By the time my colleague finished, the change in mood was evident. The people, including the leader who started out by telling him to sit down and keep his voice down, had now begun to listen more intently to what he was saying, nodding in agreement. Some stood up in his support, making gestures to leave. He urged the others to join him. In a matter of about 7–8 minutes, nearly everyone stood up, and began to walk away.

The three ladies sitting next to me stayed. They said they wanted to hear what I had come to discuss. I then announced that I would not leave so long as even one person was willing to listen. I am not sure if it was humility they sensed, or simply a result of guilt, but some people came back. Others followed, including the angry man. Only, this time, he made his stance clear by facing his back to the group.

It was my turn, now, to set the rules of conduct for those who chose to participate in the discussion. I started by clarifying my position, that I was not a government representative, that I was only here to get their opinions and suggestions on how to move forward. . . . that this area lay within the city's perimeter, and that it would eventually have to be "formalized", something that the government was planning to do anyway. . . that my role here was only to make this process viable for both sides. If they chose not to discuss with me, that was fine too; in that case, they could deal directly with the government. Some expressed discomfort with my use of the notepad, so I handed it to them, agreeing to simply listen and absorb, not write anything down. On setting the ground rules, I emphasized that each person would get a chance to speak, but on the condition that they extend the courtesy to the others to speak without being interrupted. I allowed the angry man to start, an offer which he took, still with his back to the circle of people, and still bearing the same tone. At this point, the leader asked him to leave, but this is where I interjected. We had agreed to let everyone to have his/her turn to speak, and he could have his. Moreover, I needed to hear what he had to say – he was making some very relevant points about the government's actions—both

(continued)

Box 7.4   (continued)

previous and current—which could have implications on our work on the proposed project in this community. [Part of this had to do with a road project undertaken by the national government, under which several people had been rendered homeless. The consultation process was in name only; resettlement packages had been promised but not executed properly, and so on. He feared that the same would happen here. . . . So, clearly, his sentiment was justified.] After he finished, others followed with their concerns/suggestions. The proceedings were relatively civil.

At the end, this "angry man"—who, as it turned out, was the former elected leader of the community—came and shook my hand. He said no one had been willing to hear him out until now, or clarify his doubts, and thanked me for the opportunity. I thus came away with two victories. The first was in being able to start discussions despite the "interruption", and having been able to answer many more questions that would have arisen had the discussion gone without this rigorous debate. But more importantly, the victory lay in being able to win the "angry man" over, and possibly his faction within the community. (AM)

be respected by the other if a resolution is to be achieved. A very useful way of achieving this, particularly with leaders who are more comfortable giving orders than listening, is to use role play to understand the other side. For many people this will not be easy, and we would strongly recommend that as part of the role of good government, leaders should be given training in matters such as negotiation. Role play requires people to construct an argument from the perspective of the person or organization they are in conflict with.

For example, let us take the case of a person who has a very substantial portion of land on the periphery of a settlement. He acquired it many years ago (without payment to anyone). He uses the land as a farm (employing three people) and has built a nice house on it. The community wishes to use the land for resettlement of families whose houses are being demolished under an upgrading scheme. The position could be mapped out as follows:

Figure 7.1   Upgrading scheme scenario.

Community position	His position
This person acquired the land free of charge when the area was first occupied by squatters.	When I came to this area, no one was interested in this piece of land which was outside the settlement, so what was wrong with me using it all?
He has defended his claim to the land by using force, and fencing the property.	Only after other people saw what I had done to the property, people started getting jealous and wanted to claim some of the land.
This person is being selfish by refusing to share the land with others.	I have invested a lot of money in the property, and now they want to take it away.
He has no right to use so much land when other people have to put up with tiny plots.	At least I am using the land as a business: most people don't even use the land they have.

The second point about negotiation and conflict resolution is to remove the words *right* and *wrong* and look at the situation from the point of view of what gains can be made from a settlement. The situation is always made more intractable where people have adopted a public position. If so, they will lose face if they "give in" and can be seen to be weak. They ask themselves:

- If I "give in," will I be criticized for it?
- Will I lose power and authority?
- How will my committee see me, if I am seen to be sympathizing with the other side?

It is these sorts of issues which can be the biggest stumbling block to movement in any conflict-resolution process.

At the same time, leaders may ask themselves whether their long-term reputation will be enhanced more if they have a record of resolving disputes or creating them, and whether by alienating parts of the community they are sending the right message. There is no doubt that resolving conflict is harder work in the short term than the "nonnegotiable" position; but conflicts have a way of resurfacing and, like a dormant disease, erupting into nasty sores at a later stage.

Let us look back at the land case outlined in Figure 7.1. Fisher, Kopelman, and Schneider provide an extremely good framework within which each side operates.[6] This framework is somewhat grander than the small land issue we referred to earlier, but let us illustrate the type of process which might follow in that example.

Box 7.5 Seven elements of a conflict situation.

Interests
  Have the parties explicitly understood their own interests?
  Do the parties understand each other's priorities and constraints?
Options
  Are sufficient options being generated?
  Is the process of inventing separated from the process of making commitments?
Legitimacy
  Have relevant precedents and other outside standards of fairness been considered?
  Can principles be found that are persuasive to the other side? To us?
Relationship
  What is the ability of the parties to work together?
  Is there a working relationship between their negotiators?
  Are the parties paying attention to the kind of relationship they want in the future?
Communication
  Is the way the parties communicate helping or interfering with their ability to deal constructively with the conflict?
  Are mechanisms in place to confirm that what is understood is in fact what was intended?
Commitment
  Are potential commitments well-crafted?
  Does each party know what it would like the other party to agree to?
  If the other side said yes, is it clear who would do what tomorrow morning?
Alternatives
  Does each side understand its Best Alternative to Negotiated Agreement?
  Are the negative consequences of not settling being used to bring the parties together?

Source: Roger Fisher et al, *ibid.*, p75.

1. The first thing to note is that the man *has* the land. Any attempt to remove it from him by force would result in a very bitter conflict, as well as delays. He might, for example, go to court to defend his interests, and thereby delay the matter for years. Negotiation is therefore important.
2. The next thing to note is that the landowner uses the land as an income generator; so if a new form of income generation could be devised, he might be interested.
3. Third, when the upgrading project is finished, all landowners will be paying property taxes, based on the size of their land holding. He will therefore be facing quite a substantial bill.
4. Last, he may be uncomfortable about the community pressure which is building up. He doesn't see himself as a bad guy, but is being portrayed as one.

Approaches to him could therefore start from the position that his rights over the land are fully recognized, but the community urgently needs land. Can he help? Also, he might be concerned about the property tax, which could effectively eat all the profits from his little farm. As a way out, could the community help him (with government support) to start a new land-intensive vegetable-growing farm which would yield the same income from a smaller piece of land? If he surrenders the land, he will be publicly acknowledged as a benefactor to the community. A road will be named after him.

In this type of dialogue, some people would shrink from making concessions, and/or giving him any praise for surrendering land. But it is from precisely these very small things that big changes can occur. Those who are willing to defer to the other person and give him or her respect are the ones who are most likely to get a positive response. The following quotation from the famous Dale Carnegie sums up this point:

> You want the approval of those with whom you come into contact. You want recognition of your true worth. You want a feeling that you are important in your little world. You don't want to listen to cheap, insincere flattery, but you do crave sincere appreciation. You want your friends to be, as Charles Schwab put it, "hearty in their approbation and lavish in their praise." All of us want that.
>
> So, let's obey the Golden Rule, and give unto others what we would have others give unto us.[7]

The *style* with which conflict is approached is as important as the *content*. The same book gives guidance on how to stop disagreement becoming an argument. This includes:

- Welcome the disagreement.
- Distrust your first instinctive impression.
- Control your temper.
- Listen first.
- Look for areas of agreement.
- Be honest.
- Promise to think over your opponents' ideas and study them carefully.
- Thank your opponents sincerely for their interest.
- Postpone action to give both sides time to think through the problem.

Could my opponents be right? Partly right? Is there truth or merit in their position or argument? Is my reaction one which will relieve the problem, or will it just relieve any frustration? Will my reaction drive my opponents further away or draw them closer to me? Will my reaction elevate the estimation good people have of me? Will I win or lose? What price will I have to pay if I win? If I am quiet about it, will the disagreement blow over? Is this difficult situation an opportunity for me?[8]

These attitudinal attributes can make or break any attempt to build consensus, and it is in the last paragraph of this quotation that the essence of the whole concept of successful negotiation lies. What will work for *both sides*? That is the much-clichéd win-win situation.

We would add one more very important quality: patience. Decisions which are rushed through for the sake of appearance of consensus can unravel with disarming speed. So, however tempting it may be, we do not grab at the first straw of agreement, but make sure that agreement is reached with a complete understanding by both sides of precisely what has been agreed and how it will be implemented.

We referred earlier to the need for specific skills. The community facilitator has a job which requires experience to deal with conflict in a non-directive way: this is much harder than it sounds. He or she also has to understand the technical issues involved, without purporting to become an expert. He or she must be a friend and servant of the community while honestly and conscientiously serving the local government. This is a role which we call the barefoot bureaucrat. Second, there is a need for the leadership, and the community at large, to understand and be able to manage effective conflict resolution and consensus building.

The unfortunate situation is that the skills which community facilitators—barefoot bureaucrats—require, and which they can, in turn, impart to the leadership, are offered by few conventional training institutions. Not only is there a lack of recognition that this work requires special skills, but the training technique is little understood. This is described and discussed in Chapter 8.

8   Barefoot Professionals
A New Breed of Experts

One of the most serious impediments to development and growth is that of dependency. As discussed in Chapter 5, the rules established by the formal system marginalize, and even criminalize, the disadvantaged. There are other ways in which the poor are excluded, notably in the ability to access professional help whether it be medical, legal, technical, or social. Whereas the better-off sections of society have access to a circle of qualified professionals whom they can approach for advice, and whom they can pay for their services, the poor have no such luxury. Because there is a lack of experience and understanding within the community itself of such specialized fields, they are more vulnerable to exploitation by those who pretend to have such skills. And because they cannot afford what might be called a full service, they are vulnerable to those who offer shortcuts. If we recognize that a lack of skills within the community is a serious drag on its development, how can this gap be addressed?

The message of this chapter is twofold: the first is that if government can play any role in development of housing for the poor, it must be a supportive one; and second, that support must include helping to develop appropriate skills within a community and helping them to access the resources of the state to solve their own problems.

SKILLS WHERE THEY MATTER

The term *barefoot* is used in this chapter to denote someone who is of the community and has skills specifically tailored to the needs of that community. Barefoot doctors were first used by Mao Zedong in connection with his village health program. We take this concept and apply it more widely, remembering that relevance and community linkages are the key components of the concept.

An interesting parallel in the field of medicine with Mao's barefoot doctors, especially in Africa, is provided by practitioners of traditional medicine, whom we shall call, for simplicity, traditional doctors. They use traditional herbal remedies, often with great success. There is an increasing body of work which demonstrates how successful such remedies are. These people are closely embedded within the society and share its values. They have the trust and full cooperation of the public. Their contribution to health is unquestioned, although there are, inevitably, examples of good and bad practitioners.

These are not formally trained people, but they are effective. Why? First, they share the values, language, and daily life of their patients. Second, there is a relationship of trust and respect due to social ties. Third, the treatment is affordable and practical.

It would be wishful thinking to insist that every community should have its own doctor, just as it is wishful thinking to insist that every community should have its own architect, planner, and lawyer. This is where the concept of barefoot doctors is helpful in understanding the idea of other, possibly less familiar, barefoot professionals. There are ways in which skills can be developed within the community: skills which are relevant and responsive to the needs of the community. But if so, what sort of training would be appropriate?

SCALE OF LEARNING

Donald Kirkpatrick developed a very useful model to gauge the effectiveness of learning. Applying the concepts to training events, it can be interpreted as follows:

Figure 8.1   Applying concepts to training events.

Level 1	Satisfaction	Were trainees satisfied with the quality of their training?
Level 2	Learning	Did trainees learn or acquire the knowledge, skills, or attitudes the training was intended to convey?
Level 3	Application	If they learned, did the trainees apply to their jobs, or at their workplace, the new knowledge, skills, or attitudes?
Level 4	Organizational performance	If knowledge, skills, or attitudes were applied, did that make a measurable difference to the performance of the organization concerned?

Source: USAID/G/HCD/HETS, *Monitoring Training for results* (Washington, DC: AMEX International, Inc., Creative Associates, Inc, 1996)

This table focuses on the actual usable results—making a difference. As obvious as this is, too many training courses are little more than "nice to have." There are probably three things which distinguish, for example, training experiences at the top and the bottom of the preceding scale.

- The first is whether the person wants to learn.
- The second is whether the material is appropriate in content to the needs of the trainee.
- The third is whether the material is communicated effectively, both in terms of the language and concepts used, and the pace.

Good teachers will be fully aware of the significance of each of these components in terms of knowledge transfer, and will pitch their presentations accordingly. However, difficulties arise when there are very diverse levels among the students. If the tuition is too slow, some lose interest, and if it is too quick, others give up. Balancing these competing demands is very difficult, if not impossible.

TRAINING PRINCIPLES

So, how do we train barefoot professionals?

The most important point is that the training must be *relevant*. Even though people may have an interest in a huge range of issues, in point of fact it is skills transfer that matters, and there will be a limited time frame within which the skills transfer is needed. There is no point in undertaking a training program unless there are ways in which the newly found skills can be applied. Thus a program which trains people in bricklaying can be counterproductive unless there are opportunities to lay bricks. Equally, a training program which helps people to be bookkeepers when they do not have any role in relation to accounts will not succeed, as they will quickly forget what has been taught.

On the other hand, a training program on management of storm water, in a community that is plagued by erosion and flooding from storm water, will sow the seeds for not only improving the current situation, but equally, maintaining it and steadily improving it. With this experience the people so trained can impart the skills to other communities by showing them how to do it.

The program must be *accessible*. There are many dimensions to accessibility. The first is linguistics: far too often training is undertaken in the language of government, which may be very different from that of the residents themselves, some of whom may be illiterate. To make matters worse, people typically fear that they will be despised if they cannot understand the posh language and, hence, pretend to do so. It does not need much imagination to understand how much gets lost in the process.

Second, accessibility refers to the logistics of the training. How convenient are the times of the training? Will the training times suit both sexes? Will people have to sacrifice income in order to attend? Will they have to spend money on getting to the training site? It is much too easy to assume that people will be willing to make sacrifices in order to attend training, and indeed that is not a bad thing in principle. But they should be consulted about the logistics, and solutions should be worked out in collaboration with them.

The training must be *demand driven*. If the intended beneficiaries are not really interested in what is being taught, then the whole program is a waste of time. There are many cases where courses are held and people attend in apparent willingness to learn. But if they do not really want to learn, and instead just want to get some nice-to-have knowledge: that is a different thing altogether.

There are several ways of determining whether there is a demand. Who wants to learn and what? This must be the starting point. But we cannot possibly address the learning needs of the whole community, so we must be fair to them and outline the field in which we might be able to assist, so that the discussions can take place on the same page.

The second indicator of demand is whether people are willing to make sacrifices in order to learn. We have already referred to the need to structure the training in such a way that it is practicable in terms of the scheduling and the venues. This is to reduce the unproductive sacrifices to a minimum, and establish an efficient framework with a minimum dropout rate. The other side of the coin is that unless sacrifices are made, people will not truly value the training and it will either fizzle out, due to low attendance (on longer-term training) or be viewed (as we hinted earlier) as a nice holiday.

What sacrifices work? Contributions in kind are a first step. For example, those attending take it in turns to bring food, even if simply snacks like boiled eggs, bananas, or some form of sweet. Alternatively, they can make a cash contribution, for example, for the cost of any written material which is supplied. Whichever solution is adopted, and however small the amount (in financial terms), it should be enough to help people feel a sense of commitment to themselves to get the most out of the event or the course.

It is easy to say that a course should be demand driven, but how do we identify demand? The first step is to involve all those who are affected. An excellent starting point is to look at the problem from the point of view of skills within the community. For example, how many bricklayers are there, and is there a need for more? How many bookkeepers?

Consultants flourish on skills audits and the like. They will design questionnaires, recruit interviewers, and manage the survey process, including sampling, if any. The data will be entered by specialists, and analyzed by other specialists, and much will be made of the statistical analysis. These days it is usual to hire some or all of the interviewers from within the community so that at least there is some skills transfer.

Is this methodology appropriate? Is it not a "them and us" approach? If we were to do a skills audit of our own street, would we send out interviewers? Surely not: we would convene a meeting and ask people to put their name and skills on a piece of paper. Then someone would put it all together in a list, which would be like a neighborhood advertising leaflet for local services. For example, A does dressmaking, B does cake making, C does carpentry, and D does car repair.

How?

One solution, often offered by NGOs and universities wishing to make contributions to the welfare of the poor, is to provide part-time specialist support services to communities. These may be in the field of law, planning, architecture, and so on. The concept has been successful in the legal field in respect of major cases, but is unsustainable in terms of a legal service which can address the day-to-day needs of the residents. The experience in the physical field has been less successful because the solutions lack the specificity and adaptability necessary for the poor. More importantly, this is providing a service but is not imparting skills, per se.[1]

Another approach is to ask small business people, for example, what problems they are having with their business. Are they financial, technical, management of people, or what? And if so, is there some training that they think they can benefit from? For example, an informal sector builder cannot develop his business because he does not understand estimating: appropriate training and technical support in this could transform his business.

Then there are those who have no specific skill. This is a much harder call to make. If they receive training, what opportunities will there be for employment? It is suggested here that rather than open the door of learning generally, the starting point should be to use specific problems to be addressed within the community as starting points for knowledge transfer (as in the instance referred to earlier concerning storm-water drainage).

To return to the concept of skills audit: once an audit has been done, people must decide the areas in which their skills need improvement. One part of this decision is to get more experienced people from other communities, or from NGOs or consultants, to appraise their skills and recommend the areas in which they would benefit. After that, the people must prioritize.

The next stage is to decide how training will be undertaken. We have already described the criteria which distinguish a good program from a bad one. Another way of looking at this is to understand what we expect from a training program. Too often training is done for the sake of it, and a good way of evaluating the degree of success is to look at the outcomes using the Kirkpatrick model.

Small groups are therefore ideal, and one-to-one mentoring is best. If the mentor/trainer is from the same background as the trainees—true peer-group knowledge transfer—so much the better. In that situation there is not only learning by doing but also the flexibility to respond to specific needs and interests.

Sustainability is a buzz word which has been greatly overused and misused. But in the field of training, capacity building, and local development it is absolutely key. In our view there are two prerequisites for sustainable capacity building at the community level. The first is that the community must be constantly looking at its own resources to determine what skills are needed and how they can be upgraded. The second is that they must have access to support networks.

It is on this second aspect that we must now dwell. Learning by doing implies that activity is taking place, but no one is very good at predicting what skills will be needed or whether people who think they can handle a task can actually do so. Let's take a practical example. A community has decided that they should embark on a storm-water drainage project. The objective is to solve the problem of frequent flooding and stagnant water which allows the breeding of mosquitoes. One of the community members, Lambo, is a bricklayer, and has often supervised laborers excavating trenches, building culverts, and the like. He therefore volunteers to manage the project. He is pretty clear that the job can be done in a few weekends if the people turn up in their numbers to excavate the ditches. In this way, without anyone having to pay money, the physical problem will be solved.

However, after volunteering for this managerial position, Lambo starts to have doubts. He knows that on a building site, the levels of the bottom of the trenches are set out from level profiles which have been established by a surveyor or site engineer. Could he manage without them? He knows that part of the site is very flat, so that if the trench is dug too deep, it will not drain and the situation will be as bad, if not worse than it was before. Similarly, he knows that it is impossible to tell just by looking which part of the land is lower, and if he plans the final route of the drain to the wrong side of a small hill, all the work involved could be abortive. In brief, he asks himself whether he needs a specialist to set out the levels.

Lambo decides that he cannot trust his eye, and he must establish the levels accurately. He must therefore find a theodolite or similar leveling instrument, and get the services of a surveyor. He knows he can borrow the former from a friend who works for a construction company, but the services of a highly skilled surveyor are harder to find. He needs training to be sure that he does his work correctly.

This is where the networking principle comes in. If an effective support network exists, then the community will be able to access the services of a surveyor to do the work. Or, and this is far better, to come and train the person concerned to do it.

Let us move into the scenario where the surveyor has arrived. He discusses the situation with Lambo to understand what the problem is. From then on there are two options.

The top-down option is for the surveyor to order that stakes be erected as leveling profiles, and then start with his level along the proposed route of the trenches. Within a day he has finished, the community knows how deep to dig at each point, and things are ready to start at the next weekend.

The bottom-up option takes longer.

Bottom-Up Training: The Principles

The surveyor discusses with Lambo how he would start the job: why he starts at the lowest part of the site, and how he will proceed from there. He explains about the profiles: how to erect them, how to mark them, and how to get the trench levels from them.

Next he shows Lambo the different components of the instrument, how to set it up, and how to read the numbers on the staff that another member of the community is holding. He also discusses how steep the fall must be in the drain so that they can be sure that the water will flow. Lambo has one opinion and the surveyor has another. They look at the soils and after more discussion agree on a compromise. Next the surveyor shows him how to calculate the fall on the land in terms of the rate for linear meter. They try several calculations and after a while Lambo feels that he can do it without making mistakes.

Bottom-Up Training in Practice

Lambo has great difficulty reading the instrument: everything is upside down and very confusing. He gets the levels wrong for the first few times. The surveyor does nothing, in spite of being very tempted to do the whole thing himself. He felt like a father watching his child playing a piano and torturing the music, while being so tempted to play the beautiful music himself. Sometimes Lambo will get confused and the surveyor will help him.

It takes the best part of a day before Lambo can be sure that he is doing it right. The next day he works alone. There are several times when he wishes he could ask the surveyor, but he cannot, as the surveyor has gone. On the third day, the surveyor returns and checks the work. He finds a mistake, and they discuss how the mistake occurred, and how to check for one's own mistakes. Lambo asks the surveyor about a few minor problems he had. On the afternoon of that day Lambo continues, and finds that he is now quite comfortable with the work. The surveyor returns on the fourth day, and pronounces that the work is accurate.

This is what we call sustainable training.

It is easy to suppose that training should aspire to the highest professional qualification. Today there is substantial support for the concept that all training, no matter how lowly, should be recognized as a step on the ladder of a formally recognized qualification. Thus a one-week course in bricklaying can give credits in a six-month course in bricklaying. In turn the six-month course is a module in a three-year diploma in building construction, and the three-year diploma gives credits in a bachelor's degree in building science. This is a neat and very laudable framework within which to position training programs, and it gives them legitimacy. However, at the lower levels of such training ladders, what is often being provided is essentially a cut-down version of a "proper" training program, instead of what is really needed for the purpose. There is one problem with the term *professional*—even a barefoot one. It implies a certain level of qualification. To be

Box 8.1   Barefoot planners in Swaziland.

As part of a project to upgrade informal settlements in Mbabane, Martin and Mathema proposed using community representatives in each settlement as Neighborhood Upgrading Facilitators (NUFs) to carry out mapping exercises, and serve as the liaison between the Council and the local residents. The underlying principle was to harness local ownership of the project, and maintain some level of continuity through the term of the project, after the current consultant team leaves. The NUFs were elected, or nominated by the local leaders (called Zone Leaders), to represent between 15–30 households within each settlement.

NUF Selection and Training. Following approval of the concept of NUFs by the leadership of all settlements, NUFs were elected from residents of each local group of houses. The NUFs were trained by the consultant team on topics ranging from mapping to community interaction, emphasizing their role as "facilitators". This was done to ensure consistency in the various neighborhood upgrading plans, and also to ensure that they did in fact develop their plans in a consultative and democratic manner.

Mapping Exercise. The settlement was divided into blocks following the general boundaries of the internal jurisdiction of the settlements of roughly 20–30 homesteads, each with a designated NUF. The NUFs—essentially community volunteers—were trained to carry out a mapping exercise of their respective blocks .using enlarged aerial photos of their respective areas. This included identifying plot boundaries, water points, electricity connections, commercial and institutional uses, most commonly used footpaths and roads, and so on. They were also trained to number the houses/structures by plot and section: stickers bearing the numbers were stuck to the doors of the buildings, and the corresponding numbers marked on the drawings. These numbers were later verified and used by the Census Team to carry out the formal house numbering. The mapping responsibility of the NUFs also included consultations with residents in each block, and proposing improvements in consultation with the resident communities. These proposals were developed in co-ordination with the local Leaders (and settlement leadership), to ensure that the larger objectives of the community are met, while actively taking into consideration the views of each household.

Community Consultation for Proposal Development. The NUF drawings were then translated into AutoCAD, this next iteration of existing plot boundaries and proposed improvements then presented back to them, and their feedback incorporated. The next step for them was to take this back to their communities and get agreement from each household on proposed interventions and related costs.

Participating in the Census. After completion of their mapping exercise, some of the NUFs were also trained and involved in a parallel Census mapping and house-numbering exercise.

NUF training in an informal setting

The NUFs at work, on aerial photos and maps of the settlement, plotting out the existing services and desired improvements.

clear, the concept has nothing to do with formal qualifications: it relates to serving the specific needs of the community: the insider. Box 8.1 illustrates this with a case of a small training program for barefoot town planners in Swaziland.

Barefoot Builders

Chapter 4 described how, in many respects, there is a high level of design competency in informal settlements. This derives from a great need to economize, which results in extremely cost-effective and integrated solutions, as well as expertise in a limited repertoire. Some traditional architecture has the same characteristics.

However, it would be misleading to claim that everything is perfect. There are many examples of bad building, sometimes deliberately shoddy, to save money and even defraud the customer; and at others it is simply done out of ignorance. There is a need for better standards, especially in construction, but also in terms of design, because in the process of adapting to urban life, many of the old values and systems are abandoned, and the status-seeking symbols are adopted instead of practical and efficient solutions. New materials are also used, often with unhappy consequences.

So who will be the barefoot architect/builders? Unlike the Swazi case of the barefoot planners, where skills had to be introduced which were completely new to the community, building is not a new science. There are many builders in most communities who show varying degrees of skill, understanding, and honesty. It is these people who we think can be trained as barefoot architects.

What do barefoot architect/builders need to do? Just like the barefoot doctors, they need to be able to diagnose problems and prescribe appropriate remedies, because most of their work will be improving and extending existing houses. They also, unlike their medical counterparts, will be called upon to advise on design. In this connection there are several points in which they can benefit from training.

These include design for incremental construction (how to start with a tiny but adequate working house which can be extended with minimum disruption and cost), the building regulations regarding day lighting and ventilation, the principles of structural soundness (for example, minimum wall thickness related to height and length), basic design concepts for greater resilience to earthquakes and hurricanes/cyclones, effective use of spaces and storage, and so on, heat insulation and orientation to optimize heat gain in winter and heat protection in summer. These principles can be taught over a period of several months, taking live examples, and inviting the participants to design their own solutions to problems.

In addition, they should be taught, or reminded, about good building practice. Four problems occur in building with sickening regularity.

- The foundations are inadequate: this can be a product of building on bad soils, or of not excavating deep enough or wide enough.
- The second is the incorrect use of materials. The most common is incorrect use of cement, where it is mixed up and then dries partially. More water is added so that it looks the same as it did when first mixed, but in fact it has lost most of its cohesive strength. Building blocks are also sometimes inadequate for the purpose: if they are clay bricks, this can be because they are not properly fired; if they are concrete blocks, it is usually because not enough cement was used or they were allowed to dry out before the cement had cured.
- The third is that roofs blow off. Methods for fixing roofs are well understood, but many are cumbersome for builders who prefer to use a few nails and hope that that will be sufficient.
- Lastly, there is rising damp, caused by either a lack of a damp-proof course or the external soil level being above the damp-proof course.

This sort of training is also particularly relevant from the perspective of building more safely, and reducing vulnerability to natural hazards. There are very simple, time-tested techniques for building nonengineered structures which can be applied easily and cheaply. None of these points requires a particularly long or complex training program for the builders to understand.

Lastly, there is the question of management of time and labor. There are simple project-management techniques and cost-estimating techniques which can be taught: these will help people use their labor more efficiently and thereby reduce costs and increase profits.

The value of such programs—training in design and good construction practice—will be greatly enhanced if the community plays a role in the governance aspects of the industry. It is a sad fact that in the building industry there is a need for constant consumer vigilance, and to protect consumers against fly-by-night operators[2] there must be a system by which builders are accredited in some way. The best method is for the community leadership to have a record of the work of the individual contractors. Thus prospective clients can meet the previous clients of a short list of suitable builders and see their work before commissioning them or asking them to quote.

This brings us to another area where networks are essential and the training option is less cost-effective than the network option: the twin questions of quotations and materials supply.

Small-scale contractors find it notoriously difficult to prepare accurate quotations, even for their own labor without materials. For simple jobs, elementary training can be quite effective, but the complexity of preparing quotations, especially with material prices, is beyond the normal small-scale builder.

This training could have helped one of our case study people in Chapter 3, Irene, to avoid the very common problem of starting a house which she can't finish due to lack of funds. Readers will recall that she bought a piece of land with some money she had inherited, and used the remainder to build a house. "I completed one room of the three, and the money finished." All that remains is money for windows and roofing—if she had built two rooms instead of three she would have had two rooms, one of which she might have been able to rent, and thereby raise enough money to finish the third room.

In addition to basic training in preparing cost estimates, barefoot builders can also benefit from technical support in costing and estimating. Building-materials suppliers are usually expert in such matters. Where this capacity does not exist, the training could be extended to them so that, using a hand-drawn sketch as a starting point, they can produce drawings and schedules of materials and prices within a matter of hours. This is already being done in some countries, and the software is readily available.

Pralahad[3] uses an excellent example from Mexico to demonstrate the synergistic relationship between a cement company, CEMEX, and small builders, whereby the building-material supply and delivery system was attuned to the needs of the poor. In the process, not only did they get their materials supplied on time and in good condition, but they also had a mechanism for saving for building construction. Thus, by focusing their training on existing entrepreneurs, and giving them additional skills, they help them perform a better service (to the benefit of the clients), and thereby make more money (to the benefit of themselves). They extend their services through networks, thereby giving the poor access to services and funds which would otherwise be beyond them.

Going down the chain of complexity, what about the barefoot bricklayer? There is no better example than a small project run by the American Friends Service Committee in Kafue, Zambia, where about two hundred very poor rural families got together to build themselves a house. It was a mutual aid scheme in which all families worked together on all houses, and once they were completed drew lots for which family would get which house.

The women were the bricklayers, and after a rigorous training of about two weeks, which focused on the three elements of bricklaying—setting up the vertical, setting up the horizontal, and getting the thickness of the joints between the bricks even—they were able to produce walls of astonishing quality. They built a house at a public exhibition within three days, under the very critical eye of the public. And these were people who had never before even touched a brick.[4]

Similarly, there are barefoot plumbers. They are people who specialize in handling plumbing problems within low-cost housing environments. Plastic pipes have made this environment very significantly easier because cutting and jointing are much easier. They concentrate on maintenance problems: leaky taps, WC cisterns which do not flush properly, and blocked drains. Within these fields they are specialists and masters. Do not expect them to do the plumbing in a new house, or lay new drains—that is a much higher order skill. But within the limits of maintenance they are experts. And, in that, they can be trained within a month.[5]

Barefoot Bankers

Such training and support does not have to end with the conventional butcher, baker, and candlestick maker. How about banking?

The Grameen Bank, of course, is an outstanding example of lending to the poor. For readers who are not familiar with the concept, it is to offer small loans to women's groups, and in the event of default by any one member of the group, the remainder are liable for repayment of that debt. The groups are typically about five people. The results have been so successful that the concept has been copied in various ways throughout the world, and it is the foremost example of banking for the poor, though in a very limited sense.[6]

Most of such small-scale lending programs are managed by nonbank financial institutions, typically NGOs, which evade the restrictions of registering as a deposit-taking institution by only dealing with loans and not savings. This somewhat distorts the relationship: we think that savings are at least as important as loans, and play a very important part in developing a community's sense of worth. The savings habit, however small in terms of regular contributions, is the foundation for good financial management.

We therefore think that training in money management, associated with a community-based savings scheme, is a very important tool in the economic growth of a community. The example of HFC Bank from Ghana, which was described in the last chapter, shows what can be done by people who, on the face of it, are very poor and have few prospects of growing out of poverty.

THE BAREFOOT BUREAUCRAT

It is odd that though the concept of governments as enablers has been with us since the 1976 Habitat Conference in Vancouver, little progress has been made in restructuring government services to perform that task. The situation deteriorated in the 1980s and 1990s when the Reagan-Thatcher antigovernment sentiment effectively made government a dirty word. The time has come to relook at what governments can and should do.

For a poor person who has the opportunity to build a house, what services would he or she expect? And, looking at it from the public policy point of view, how would government's services be structured so as to expedite delivery?

This is where the barefoot bureaucrat (BB) comes in. The BB is there to help people steer their way through government controls and to access government supports. For example, he or she will help the individual complete formalities such a having house plans prepared, obtaining building permission, obtaining loans or credit with building-materials suppliers, getting materials delivered to the site, engaging a barefoot architect, and so on.

In this the BB acts as both a mentor and technical advisor. He or she is someone to whom the ordinary people can turn when things go wrong, someone to fight for them and make things happen. This is a total reversal of the role of the more typical bureaucrats who use position to show power over other people, enjoy saying no, and who use forms and procedures to demonstrate their superiority. What is the difference between a barefoot bureaucrat and a normal one? The normal bureaucrat is there to represent his institution, to follow the rules, and to do as he or she is told by his superiors. The face which the normal bureaucrat presents is obstructive, obscure, and arrogant. By contrast, the barefoot bureaucrat is there to serve the community in which he or she is working, and facilitate the maximum benefits from his employers on behalf of the community.

The BB therefore represents a reversal of government's typical role from controller to supporter.

Support and Assistance

The function of a BB must be determined by what the poor person needs. Just as a wealthy person hires a lawyer to deal with legal disputes, and gets an architect to handle the whole process of designing and constructing a house, so the poor person needs support. Governments cannot pay architect's fees, but they can help people to design a house, select a builder, and obtain materials. These support services can be provided quite cheaply and effectively if the bureaucracy is structured correctly. This is *good government* in contrast with the bloated, self-serving civil service which was the old-style *bad government*.

We must beware of clichés, but the BB is there to give power to the people to solve their own problems. Included in this will be, for example, the means to make an informed decision. Box 8.2 describes one way in which this has been successful. These decisions affect how much the householder will pay and how much space he or she will have to live in.

Box 8.2 The ability to choose: Barefoot Bureaucrats in action.

In South Africa we were able to help communities decide their priorities in terms of housing solutions. By modeling the trade-off between (principally) infrastructure standards (especially sanitation and roads), plot size and house size, rapid and effective decisions were taken. Beneficiaries were divided into groups of about 20, each of which would occupy one portion of the site. The first stage was to agree how much they could afford to spend per month on housing and services. After that they took their own decisions on the standards they wanted to use. This effectively overcame the mismatch between costs and income. This system was one which no community could have developed for themselves, and it was a contribution which, in normal circumstances, the state should make to assist the users to design a solution that they could afford, and which met their life-style needs.

As an illustration of this, the capital cost of the following two solutions is approximately the same. The first is the standard minimum standard for infrastructure in South African low cost housing, the second is the solution typically chosen by ordinary people. To put the matter into context, the first solution is that typically proposed by engineers: it has high standard roads and waterborne sanitation. The cost of waterborne sanitation includes sewers as well as the fittings (as well as, of course, payment for additional water consumption and sewage treatment). The house is on a 12m x 24m plot, with 5m tarred roads, in 8m road reserves, and 16m distributor roads.

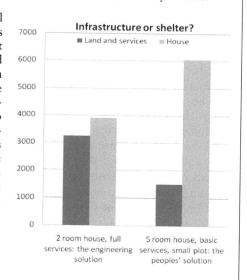

The second solution is a smaller plot (8m x 16m), and the house is semi-detached. Access is by footpath, with graveled distributor roads. The graph shows how minimizing the cost on roads and sanitation enables the construction of three additional rooms. The ratio of shelter to infrastructure on the first solution is 1:1.2, in the second case it is 1:4.0.

Training Barefoot Bureaucrats

This use of BBs has management implications which are discussed in the final chapter. At this stage we must look at the training requirements of a barefoot bureaucrat.

As we have already described, the BB has a job which requires experience to deal with conflict in a nondirective way: this is much harder than it sounds. He or she also has to understand the technical issues involved, without purporting to become an expert. He or she must be a friend and servant of the community while honestly and conscientiously serving the local government.

Second, there is a need for the leadership, and the community at large, to understand and be able to manage effective conflict resolution and consensus building. The unfortunate situation is that the skills which community facilitators require and which they can, in turn, impart to the leadership are offered by few conventional training institutions. Not only are the qualifications, in themselves, not recognized in many countries, but how to train is little understood.

We can get some insight into this by looking at a case study of the training program developed in Zambia in 1974. Zambia had a long tradition of community development, and an established, government community development training college. Their graduates normally went in to government service and acted as development stimulators in the rural areas, for example, helping villagers to undertake self-help projects such as road and water schemes, small business development, and so on.

When the Lusaka Squatter Upgrading and Sites and Services Project was designed, it was calculated that, in order to achieve the level of community participation required, 50 community development workers would be required. Since the national reservoir of such staff was very small, and they were used to working in the rural areas, it was decided to train all the recruits from scratch. A more appropriate method of training, which combined a modern residential six-month training course followed by six months of field training, was devised.

The basis for the training was that community participation requires people to be able to deal with many points of view, to deal with conflicts, to be honest with themselves and with others. The tool used for this was the so-called "T-group" training in which the participants meet in small groups without any programmed purpose. The course was designed and run by foreign consultants, based on well tried and tested methods from the National Training Laboratory in the United States and the Tavistock Institute in Britain.

Because many readers will not be familiar with the concept of the T-group, we think it is worth describing it briefly.

A typical T-group might begin with the "trainer" saying that he imagines that the group members have come to learn how people behave in groups by learning from their own experience of becoming group members. He offers neither agenda nor any suggestions as to how to proceed. In the vacuum that is created, group members begin to act in characteristic ways: some may try to suggest structure by the formulation of an agenda or the selection of a chairman, others may complain about the leader's failure to lead, while others may comment about the anxiety which the lack of structure creates in them. These comments are likely to be interspersed with periods of silence. Members experience considerable tension as they begin to cope with the ambiguity of the situation . . . The group looks to the trainer for guidelines, and the trainer in turn reflects back the group's desire for guidelines. Finally a member suggests some easily agreed-to action such as all the members introducing themselves and the group, seizing on it, agrees to proceed in this manner. The first session typically ends with members feeling confused and bewildered about what happened during the session and about how they ought to proceed next.

The succeeding few sessions provide further evidence of group members' efforts to cope with the frustration of ambiguity and lack of structure. Abortive attempts at leadership develop; the group may become divided into two camps—those who very much want leadership and structure and those who are opposed to organizing until the group decides what it wants to do. The trainer's suggestion that the members explore their feelings of frustration at this inability to get the group going is typically ignored. There may be a variety of other abortive attempts to organize, set up committees, elect chairmen; and the sense of frustration grows. Eventually the group responds to the trainer's comments that it will examine the member's contributions to the lack of group progress.

At this point focus shifts to an examination of member behavior and interpersonal style. Attention might be first directed to a member who has been particularly active in an earlier effort to elect a chairman. . . . The mutual support members have given each other during the period of feedback may lead to a great deal of expression of positive feeling between members; for example Don may be lavishly praised for his willingness to take a good hard look at himself. Group cohesiveness is now at its height . . . The concern of the group may then turn to comparing individual members in terms of their degree of involvement and efforts are made to bring in those who seem insufficiently involved. Ned is again focused in an effort to make him feel more a part of the group. As the time draws to a close, attention is turned to evaluation and examination of what the members have gained from the group both in terms of their impact on others, and knowledge about how the group process works.[7]

In the process of dealing with the cognitive dissonance created by the uncertainty of the T-group, and the irritations of other people's behavior in such a tight-knit social setting, the trainees acquired the skills to deal with the vagaries, unpredictability, anger, and frustrations of public participation. It gave them the inner strength to remain calm and supportive throughout a process that inevitably leads to conflict and even personal attacks.

The results of this training program were very remarkable. Although the process itself was criticized (for example, by the staff of the Government Community Development Training College) as bringing new values into Zambian culture, and being unsuitable to the values of an African country, the performance of the graduates was outstanding. The classroom training gave them the personal strengths, while the field training gave them the understanding of the systems being used in the upgrading program. Many of them rose rapidly in the hierarchy, and after the completion of the project went on to positions of authority.

People with such a background can also manage training programs for leaders in terms of how to manage people, conflict, meetings, and so on. The issues which we described earlier would feature heavily in such training. But the training would not be all: the community facilitators themselves would be available to assist when required.

Training is not enough. There are many ways in which something more tangible is required. This chapter has restricted itself to the need for and nature of training. How that training is applied, the role of the different stakeholders, and the interface between government and people in typical upgrading or self-build projects are the subject of the final chapter.

9 Fair Trade
Where Economics and Finance Make a Difference

Economically deprived or bursting with potential? There are many contradictory forces affecting the economies of informal settlements. This chapter looks at new ideas to help small businesses in informal settlements to capitalize on their strengths, and using new technologies and networks grow their markets and skills.

The easy way to claim to have "developed" a settlement is to have spent money on improving the infrastructure, because spending money is a comfortable top-down process in which experts have control. Far harder is to facilitate economic growth by individual families and their community as a whole. Indeed, it is probably true to say that economic growth is the foundation of true development, as poverty is the most acute problem being faced by ordinary people.

Our case studies in Chapter 3 showed how some people have, through hard work and careful financial management, succeeded in accumulating assets. Florence is a good example: working her way up the income chain she deliberately set out to "make money." By choosing to live alone, while her husband and children remain in the village, she had limited her expenditure and competing demands on her time. She graduated from working as a shop assistant and a porter (saving a little money every day) to the stage where she could buy a small piece of land and build a house from which she could operate a food business. She now has three employees and a monthly income of about $200 per month. This level of self-sacrifice is exceptional and has its reward, but cannot and should not be the norm. For the few Florences, there are many who only manage to scrape together a subsistence income. Is there not a way in which entrepreneurship can be supported more effectively?

The development path in this area is littered with well-intentioned but ultimately unsuccessful efforts. A common example in our experience is the formalization of informal market areas. This typically takes the form of "properly" developed shops and stalls which, while being neater than the informal equivalent, act as a drain on the individual entrepreneur who must now pay rent. In other words, they reduce the viability of the enterprise rather than enhance it. The situation is also typically worsened by the fact that these new facilities are in a different location (where they will not spoil the appearance of the city): this location being out of sight will also be out of mind, and the traders/manufacturers will lose all but the most determined customer. Thus it is that bureaucrats, well intentioned or not, distort the market and kill enterprise.

Is there another way of approaching the problem? We write as the world gasps in horror at the absurdities of a subprime mortgage system that has destabilized the financial system and revealed the greed and perverse incentives of those who operate it. As banks are forced to appeal to governments for help and recessions hit the world's leading economies, we must ask whether there are other ways of working.

Interestingly enough, many of the ideas which have emerged in the last twenty years to stimulate economic growth at the bottom end are both sounder than the sophisticated balancing acts of the trading floor and, thankfully, simpler. We do not purport to be economists, but can observe what works and what does not, and draw some tentative conclusions.

MICROFINANCE

Money is a primary driver in most people's lives. From the poor in the streets of Accra or Mumbai selling a few shoelaces or packets of salt, to the overblown hedge-fund trader, money matters. But money also has a negative ring to it: the excesses of capitalism and the harm done to the poor by exploitative enterprises have given moneymaking a bad name. As the gap between rich and poor has widened, we have watched helplessly as millions more are condemned to poverty.

As discussed earlier, the role that the HFC Bank played in Ghana is supporting market women has made a breakthrough in terms of helping them to accumulate savings. In terms of lending, the Grameen Bank[1] took a pioneering leap into the dark in lending to the poor: an example which has been replicated with remarkable success throughout the world.

The fundamentals of microfinance are the opposite of the more advanced financial markets whose complexities, in some sense, led to the subprime market. The two principles in microfinance are that the lender and the borrower are linked by personal trust, and the consequences of default are felt immediately. Second, security lies in personal guarantees (usually by lending to people in groups) rather than relying solely on assets as security.

Microfinance emerged, initially, as largely an NGO activity, usually capitalized with donor-funded grants or soft loans. With time, however, it can become completely self-sustaining—borrowing capital on the market and on-lending at scale. For example, in Morocco, in 2007, the microfinance industry had 1,045,215 loans outstanding, employed a total of 4,327 personnel, and had a loan book of about $435 billion. Of the two biggest institutions, Al Amana ranks 15th in the world in terms of the number of active loans, while Zakoura is 19th in the world. They raise all their capital on the local markets and on-lend to their customers.[2]

There is no simple explanation for the huge impact of microfinance. In one sense, short-term loans are only a substitute for savings, and their only economic benefit is to bring forward purchasing power by the term of the loan—say three to six months. From the economic point of view this is hardly going to make a major impact. However, there are side benefits which might be more significant than this mere transfer of funds.

First, there is the development of discipline in financial management: this includes regular savings and/or loan repayments. But it also includes the discipline to focus expenditure where it really will make a difference, or, to put it in more technical terms, to transfer spending from consumption to production. Even very poor people have the temptations to spend on "nice to have" goods and services; microfinance gives them the discipline and determination to defer that gratification until they have adequate funds to do so.

Second, there is the self-respect associated with successful financial management and budgeting. Taking control of one's finances, and being recognized for so doing, is very liberating and empowering.

Ultimately, though, the biggest impact is that it is a mechanism by which capital can be accessed. For every unit of capital employed, there is the potential to increase income. The historical situation for the poor—which stretches back to the beginning of commercial banking many centuries ago—has been that because they lack the assets which can be used as security for a loan, they cannot raise capital and are therefore unable to break away from their present position as exploited employees or subsistence farmers. The microfinance movement has changed that. In our own work in the field of housing it is astonishing to see how microfinance has enabled the accumulation of capital.[3] The usual formula is to borrow short-term money (3–5 years) to build a room which is then rented out.[4] The income from that investment is then used to repay the loan. When the loan is repaid, a second loan is taken out, and so on. In this way substantial assets can be accumulated.

GROWING ASSETS

One of the lesser understood aspects of development has been how to stimulate local economies, and especially the role of the state in that matter. The sad tale of public officials who believe that pumping money into facilities is the solution is repeated too often. The facilities may present good photo opportunities as the mayor performs the opening ceremony. But too often people neither need nor can afford to pay for these white elephants. Instead, much more creativity needs to be applied to both the economics of the situation and the design and placing of the products or services.

As an example, let us look in more detail at the economic situation of an urban informal settlement in a large city. At present, say, 30 percent of the community members have formal jobs. Although they bring in money, they often do much of their shopping in the mainstream shops which are cheaper than the small kiosks prevalent within the community. About another 20 percent are self-employed in trading (some have shops; some are mobile hawkers), and about 10 percent are manufacturers: carpenters, metal smiths, and tailors. A further 10 percent provide services such as hairdressing, telephone services, food production, and laundry. The remainder are either out of work, or earn so little that they are on the margins of survival. Of the 70 percent whose basic source of income is the community itself, there is little prospect for growth, as their clients are themselves poor.

There is a way out of this poverty trap. What needs to be done is:

1. To bring money into the community.
2. To increase the rate of return on labor.
3. To maximize the income of local residents by ensuring that as far as possible they spend their money within the community.
4. To increase employment in manufacturing, and expand markets.

Bringing Money In

Until they can market themselves and their products to a wider world, there will be little change. But here is the difference: the Internet has revolutionized trading, and now the poor person in Bangladesh, for example, can establish a presence which was previously unthinkable. Also, the fair-trade movement, especially the work of Anita Roddick, whose Body Shop empire made a point of dealing fairly with poor suppliers from developing countries, has given small and remote suppliers a new opportunity.

But the world is a competitive place, and neither the Internet nor fair trade per se offers guaranteed income. The product must be well designed and well made: although well-designed and well-made products take very little extra effort, they command a much better price. And, whereas previously breaking into the market was very difficult, if not impossible, today the Internet and a more flexible marketing structure make it a reality.

Let us construct two scenarios to show how this will affect a community of one million people. The first scenario assumes that the population is living at subsistence level and cannot afford to save. It will also be noted how, due to the lack of savings, the capital base is eroded due to depreciation (set at a modest 2 percent per year). The situation is exacerbated by an increase in population, thus reducing the amount of capital per head over time.

Capital 1 per person	Capital per unit of production	Annual Income	Total prod.	Depreciation after 10 years	Annual savings	Total GNP	Total pop (start)	Total pop (end)	Capital per person (end)
$900	$3 capital/ $1 annual production	$1 M	$300 per person = $300 M	$835,000	$0	$300 M	1 M	1.2 M	$628

The second scenario assumes that with good design and correct marketing, and a higher proportion of the residents engaged in manufacturing, the annual income rises from $300 to $450 per person, while the capital employed stays the same.[5] The model assumes that the residents will use some of their increased income as savings, and thereby slowly accumulate capital.

Capital 1 per person	Capital per unit of production	Annual Income	Total prod.	Depreciation after 10 years	Annual Savings (30% of diff. between $300 and $450)	Total GNP	Total pop (start)	Total pop (end)	Capital per person (end)
$900	$3 capital/ $1.5 annual production	$1.5 M	$450 per person = $450 M	$835,000	$50	$450 M	1 M	1.2 M	$1,170

There are interesting examples of how the more imaginative formal sector businesses have begun to learn the value of serving informal settlements. By so doing they are, of course, creating assets within the community and increasing employment.

Neuwirth tells how a bank in Brazil opened a branch within the *favela* of Rocinha: they made a special offer of credit cards to the residents. Initially the limit was $80, but the cards were a big success. For the residents they offered a hugely important step on the road to respectability. Similarly, the largest film-developing company in Brazil, with 160 franchises, opened a branch in the *favela*. It was the first store in Rocinha to have air-conditioning, and represented both convenience and status for the residents. Even McDonald's opened there, if only as a kiosk.[6]

Increasing the Rate of Return on Labor

Capital can be used to buy machinery and materials, the raw ingredients of production. The intermediate technology movement started by

Schumacher is devoted to making machinery which is appropriate and affordable for small producers. Polak is a more contemporary advocate of similar ideas, and has demonstrated how appropriately designed drip irrigation can be used to multiply peasant farmers' incomes many times over. He notes how the rate of return for the poor is potentially higher due to their low overheads.

> If small-plot farmers can take advantage of their low-cost labor to grow labor intensive, high value cash crops for the upscale market, why can't slum dwellers become the beneficiaries of outsourcing opportunities? Why can't they take advantage of their low labor rates to produce high-value, high margin, handmade good that sell for a much higher price than what these workers now make?[7]

Land is, in many informal settlements, already available: there is thus often a relative advantage for informal settlement dwellers who do not have land costs or property taxes to pay. Dharavi in Mumbai presents a vivid example of the advantages thus offered in terms of business efficiency (see Box 9.1).

Spending Money Within the Community

Money revolves, and the faster it revolves the more employment is created. Each unit of currency spent has the potential to generate other spending. This self-evident truth is one which many informal settlements are not organized to make work for them. Clearly there will be goods and services which they cannot produce, or which it would make no economic sense for them to produce. However, there are many opportunities for mutual trade which can be maximized. The case studies in Chapter 3 showed how enterprising some residents can be in responding to the needs of what is, indubitably, a limited market. Readers may remember the case of Kofi. He noticed that the women working in a nearby market had no toilet facilities. The illegal status of the market meant that the municipality would not be likely to provide these either. Thus he identified not only a need within the community for sanitation—toilets and showers—but also a commercial opportunity. He had to fight for approval to build the structures, but having done so was making a very good income from his initiative. This example shows how even satisfaction of basic needs has a commercial potential, and by residents taking the initiative, the money can be kept within the community.

Demand, Supply, and Serving the Market

Those of us who have undertaken social surveys in informal settlements, and among the rural poor, will have been struck by the very limited grasp that most people have of the potential value of their work,

Box 9.1 Mumbai's Dharavi "slum": A wasteland or an economic hub?

Estimates of the population of Dharavi vary from 350,000 to 600,000. Those who have never ventured into Dharavi may imagine it as a wasteland of tent-like temporary structures, an immense junkyard crowded with undernourished people completely disconnected from the rest of the world, surviving on charity and pulling the economy backward. Beneath the sea of corrugated tin roofs, the reality could hardly be more different. Dharavi is a highly developed urban area composed of distinct neighborhoods and bustling with economic activity that is integrated socially, economically and culturally.

One of the important characteristics of Dharavi is the predominance of live-work arrangements. Walking through any of the narrow residential streets of Dharavi by day, one can observe the intense productive activity taking place in nearly every home. Within its 223 hectares are about 5,000 industrial units with a wide range of industries and other enterprises including recycling, paper, plastic, steel, old refrigerators, textile, pottery, leather, stitching, printing, eatables, and restaurants. The estimated annual value of goods produced in Dharavi is $500 million ("Inside the Slums," The Economist, 27/1/05).

Dharavi is famous for its leather production. Skins of slaughtered cows, goats and sheep from all over Mumbai are collected here, from where they are sent to Madras for cleaning, then brought back to Dharavi for processing, and coloring, and shipped worldwide. This leather is notably used for Reebok and Adidas shoes. Similarly, the embroidery industry in Dharavi produces labels for brand-name products such as Pepe Jeans.

Production and manufacturing in Dharavi, however, is not an integrated process as in Chinese factories. Instead, all producers fix their own price as a function of the demand and the cost of materials. This means that a range of industrial goods are produced in small live-work units. This model, based on a multiplicity of independent producers, makes the production process extremely flexible and adaptable. The unplanned and spontaneous development of Dharavi has led to the emergence of a particular economic model characterized by a decentralized production process relying principally on temporary work and self-employment. The fact that many goods produced in Dharavi are sold on the national and international markets proves the viability of this system.

According to Sheela Patel , Dharavi "probably contributes far more to the Indian economy than most special economic zones. It also provides incomes and livelihoods for hundreds of thousands of Mumbai citizens who would otherwise have no employment. It also provides cheap accommodation. Conditions may be poor and most housing is very overcrowded but Dharavi is one of the few central locations in Mumbai with cheap accommodation – even if this is renting a bed in a room shared with many others."

Prince Charles, who visited Dharavi in 2003, cited Dharavi as a model for environmentally and socially sustainable settlement because of the way it was organized around people's needs. He was struck by what he described as the "underlying intuitive grammar of design" that, he said, was "totally absent from the faceless slabs that are still being built around the world to 'warehouse' the poor."

Sources: 1. Katia Savchuk, Matias Echanove & Rahul Srivastava. " Intro: Lakhs of Residents, Billions of Dollars" (http://www.dharavi.org/); 2. Sheela Patel and Jockin Arputham. "An offer of partnership or a promise of conflict in Dharavi, Mumbai?", Environment and Urbanization Vol. 19 (2), October 2007. (cited from www.dharavi.org); 3. Hasan Suroor. "Slums better equipped for challenges: Charles," *The Hindu*, February 7, 2009.

or the variety of opportunities that exist in practice. Just as middle-class families aspire for their children to be doctors and accountants, so the poor aspire to open a shop, be a tailor or dressmaker, carpenter or mechanic.

The ladder of their aspirations not only ends quite low down, but also can hit quite unrewarding spots. For example, the ubiquitous corner shop in informal settlements can be a nightmare to run. Due to the price and choice competition of supermarkets, it will only make a profit by selling convenience goods in small quantities and when other shops are closed. Price competition between similar shops in the neighborhood ensures that no one makes a good living.

In terms of trade, it is often sad to see the level at which the potential for goods which have a high labor content, such as wooden furniture, is not seized because there is no understanding of the demands of the market. Also, due to the lack of a modest investment in facilities (for example, having a level concrete floor on which to work), or an electric plane/drill/saw, how the quality of the work is such that the investment of time and materials falls far below the potential. The tremendous contribution that informality makes is the provision of free land and low-cost facilities. This can be leveraged to achieve much higher profits if the right markets can be found.

Several trends are emerging to help such poor entrepreneurs to achieve their potential productivity and profitability. These include such systems as bringing in international designers for the products and linking producers to marketing opportunities. This has been very successful in the world of crafts: the amount of work which goes into a well-designed product is the same as that which goes into badly designed (tourist schlock) stuff, but the price it sells at is far higher. Box 9.2 provides an exmple of tennis rackets being produced in an informal settlement in Bangkok.

In the field of furniture, it is really sad to see badly finished furniture (of grotesquely antiquated Victorian design) being lined up for sale along the paths of an informal settlement. The motivation for the design is

Box 9.2   Tennis rackets in Bangkok.

In Klong Toey, in a leaky wooden shed, set on a canal of filthy black water, I came across a tennis racket factory. It was staffed by four men. They used simple metal forms to shape the wooden frames, locate the holes for the stringing and so on. The finished frames were beautifully varnished. The red and gold lettering announced the make, and the fact that the King of Thailand used the very same racket.

There was no way of knowing that they had been made in such circumstances, and they sold throughout Thailand in all the good sports shops. (RM)

clear: the furniture offers an aspirational symbol for a family hoping to enter the world of the affluent. Its very scale and the elaborate curlicues speak of a lifestyle which only a few can aspire to. That's fine: there is nothing wrong with either bad taste or status symbols. However, the furniture is the product of two problems which the makers experience. The first is that, in order to keep the product cheap, they do not spend much time on the finer aspects of finishing; the second is that they are only serving their local market. If they knew what the market as a whole required, they might do it very differently.

In a slightly different way, we revert to the work of Anita Roddick of the Body Shop, which is a good example of how international markets can operate: she would locate suppliers of materials which would be used for the manufacture of her products, and gave them not only continuity of income, but a fair share of the price which customers would pay. Similarly, the fair-trade movement in commodities such as coffee is another example of the top-down market-based solution where the terms of trade are not biased against the producers.

Thus when we are talking about economic growth for informal settlements, it is clear that linkage of the producers and labor to markets is one of the fundamentals. There is tremendous potential for growth and development.

PRODUCTIVITY AND TOOLS

Productivity is one of the keys to economic growth, and there are many wonderful examples of how productivity can be enhanced in agriculture—for example, by the use of irrigation, fertilization, different crops, and different seeds. In informal settlements there are multiple examples of low productivity. Some, as we have pointed out earlier, are due to a poor understanding of the market, and therefore production of a small number of items which sell very slowly. Others are due to a lack of equipment.

But if there is one basic stimulus to productivity, it is electricity. The quantum leap that it enables due to the use of machinery cannot be over-estimated. Linked to a system for financing the purchase of equipment, electricity enables a completely new way of working.

However, sometimes people neither know how to use or maintain tools nor what to make and for whom. In this situation, outsourcing can be a useful bridge. The model of the 80s and 90s which had widespread application of outsourcing, especially in Southeast Asia, has further potential to be developed within the communities themselves: instead of constructing large centralized facilities, the homes and yards of dwellers can be used for manufacturing. Better understanding of economics can help accelerate development.

The value of outsourcing, if done ethically, is that the individual "manu-facturer" is given the support and assistance to make products to a high standard. Technical assistance and advice are provided, and an enlightened outsourcer will help its labor force to diversify. The big advantage of this system is that the materials are usually supplied, so less money is required by the operator. However, there are also examples of outsourcing being grossly exploitative—we must not overlook the risks.

India has trodden this path, and there are multiple examples of these "hutment" factories, making a variety of clothes, toys, and similar goods: Dharavi in Mumbai, presented earlier in Box 9.1, is a good example. There are also many examples from China of exploitative arrangements.

As technology advances, recycling is another area which is yielding increasing benefits for the poor, not just from their own produce but that of the city as a whole.

PRIVATE-SECTOR SUPPORT

This brings us to consideration of the role of the formal private sector as a whole. How can the private sector work with the poor and disadvan-taged? There are several very important linkages to be made. Although the poor have modest resources, there are very many of them: therefore they do represent a substantial market.

As mentioned in the previous chapter, a cement company in Mexico, CEMEX, offers a good example of addressing the needs of the poor while also doing business. Poor people have long realized that when they put their savings into a bank, they are effectively losing money as inflation is often higher than the interest earned. One category of goods which seems consistently to have increased in price faster than inflation is building materials. That is why people often save in the form of build-ing materials, for example, buying one hundred concrete blocks a month, or sand, or door frames. The trouble is that they have nowhere to store the materials apart from their house in the open air. By the time they come to use these materials, they have usually deteriorated substantially due to the wind and rain, pilferage, and so on. Another problem they faced is the difficulty of organizing transport for materials. The small loads which they might be able to afford at any one time are uneconomic for a normal truck.

The company therefore opens savings schemes for the residents in infor-mal settlements under which cash savings are translated into materials.[8] Thus a saver produces a list of what he or she wanted to buy, and the sav-ings are put into those specific commodities, which can be delivered to the saver's site any time he or she wishes. This works for both parties: it gives the savers confidence and stimulates their savings, while bringing substan-tial trade to the company.

Similar support can be provided to small builders. They take their rough drawings to the building-materials supplier, who then transfers it to a computerized program which:

- Prints out the drawing with all the details.
- Calculates the materials needed.
- Calculates the manpower needed for each element of construction.

This will help the builder to price his services to the client as well as manage his labor effectively. It reduces the risk of over- or under-ordering, and helps him to control the extravagant or careless use of materials. As with CEMEX's building-materials savings scheme, this obviously works for the building-materials suppliers who get trade out of it. A system which the materials yard might use in order to reduce abuse of the system is that the drawings and materials schedules are produced for a small fee which is refundable when the materials are supplied.

A less common example, and in some respects a very controversial one, is the potential partnership between communities and utility companies. For the sake of argument let us assume here that they are private companies, which is, of course, not the norm. (The arguments against private-sector involvement in public services are well known, but that debate can be held elsewhere for now.) But supposing that the companies are private, they will have access to capital, which is often not the case in public-sector utilities. In such a case, the water company, for example, may choose to enter into an agreement with the community to develop pipelines to serve the community, provided they agree to make payments for water consumed, for example, through a system of prepaid water meters. In this way a new service is provided which offers a substantial gain in convenience for the residents, and, if international norms prevail, will also reduce the cost of water. Meanwhile, the challenge of raising capital to finance the water pipes has been addressed.

There have been many clever examples of using the private sector to support individual and community development. For example, the municipality will award a contract to a private company to remove garbage. Community members with experience as truck drivers are then invited to compete to become drivers of the trucks. The drivers buy the trucks on an instalment system during the contract from their income. Until they price is paid in full the trucks are maintained by the seller, the private company. Community members also act as the collectors, and others as street sweepers. In this way, by the end of the contract, all capital and skills have been passed to the community members.

TARGETED SUBSIDIES

Utilities offer an area for a different type of stimulus. Kofi's enterprise, discussed earlier, showed that there is, in many communities, a massive demand for basic services. These are often being provided inefficiently and expensively. Good infrastructure, especially roads, water, and electricity, have a multiplier effect on economies and therefore offer a fertile ground for economic stimulus.

In this context, it is worth considering introducing subsidies based on the communities' signal that they are willing to maintain assets. This can be an effective tool for stimulating local enterprise in the provision and management of infrastructure. This is particularly true in the field of toilets and water, where running costs are everything.

This type of partnership between government and communities has many other implications which are dealt with in more detail in Chapter 11.

TECHNOLOGY AND TRANSACTIONS

The seeds for a new way of doing things are being sown. The disadvantage suffered by the poor of having no capital and insufficient income to be able to afford formal trading or manufacturing premises is being reduced by new relationships made possible by the information age. In brief, there are more poor-friendly tools than there used to be.

It is worth looking at a few examples of how the information age has the potential to transform the way that the poor do business.

Expanding the Market

The power of the Internet has yet to make much impact on the poor. But there is no doubt that there is tremendous potential for producers to collaborate and market their products on the Net. If one recognizes the goodwill which exists in many people of the so-called West, then not only are producers going to be able to increase their market but also reduce their costs by eliminating some of the links in the marketing chain. Small craft items, such as necklaces and baskets, cloth, pots, belts, and handbags, and so on, for example, are all suitable for this type of marketing provided that good photographs are available.

Reducing Transaction Costs

One of the most interesting developments, which is in a trial stage in South Africa, uses cell phones as a tool for banking in relation to local shops, thus reducing transactional costs for both deposits and withdrawals to about R0.50 (approximately $0.07). The system works as follows: Account holders, who are signed up by sales agents in their own neighborhood, receive a debit card and a cell-phone Simcard, either of which could be used to perform core banking transactions. A local shop takes on the role of an agent for the bank, where clients withdraw money, deposit money, or make a cashless purchase.

Unlike the usual banking transaction, which goes through the national payment system and requires overnight settlement, these transactions are settled by MobileMoney's own infrastructure, which sit on the cell-phone company's network. The transaction happens in the time it takes to send a text message and get a reply. For instance, as soon as a client deposits money at the shop, a text message arrives updating his or her account balance. The same goes for purchases, cash withdrawals, balance inquiries, or the purchase of prepaid electricity or airtime. Retailers use their own takings to make cash payouts, although the size of a withdrawal is limited by the amount of money the shop has in the till. The retailer is backed up by the Standard Bank brand. The service is identified by the display of the Standard Bank logo on the shop wall.[9]

Efficiency in Financial Transactions

Technology also allows people to bank and make money transfers without having to queue. One of the biggest problems for poor people used to be the need to queue for hours in order to deposit of withdraw money at a formal bank. This waste of time is now becoming a matter of the past with the increased use of ATM cards and cell-phone technology.

The examples are becoming well known: how rural women in Bangladesh can pull themselves out of poverty by borrowing to buy a cell phone and selling the air time; and how microfinance can operate even in remote areas by the use of cell-phone technology. Indeed, cell-phone technology has become one of the most important levelers of today, putting, as it does, power into the hands of everyone alike, rich or poor, rural or urban. Within the next few years we can expect a massive increase in its use as more than a simple communication tool.

These then are the principles. The last chapter illustrates how they will work in practice, and gives some insights into the factors which contribute to success in such matters.

10 Who Did What?
Monitoring, Evaluation, and Corruption

Evaluation has established itself as an essential component in the development lexicon. From the parliaments and boardrooms of funders to government offices, from multilateral donors to NGOs, and from the classrooms of top universities to the tabloid press, everyone demands to know how money was spent, whether project objectives were met, and whether the project was implemented as planned. These questions are legitimate, but is the methodology used appropriate, and are the right questions being asked at the right time? We propose that much more emphasis should be placed on local control and addressing problems as they occur rather than waiting for findings of evaluations which often occur many years later.

INADEQUACIES IN CONVENTIONAL EVALUATION

How do we know whether we are getting good value for money? How do we know whether the money was even spent on the project or the activities it was designed to perform? And how do we identify flaws in the design and implementation of projects? Simple: we set up a rigorous project monitoring and evaluation system.

Or is it so simple? In this chapter we look at how these instruments are used and whether there might not be a better way of performing the same functions. It questions whether the conventional systems are designed to benefit the beneficiaries or to merely serve the reporting needs of the funders. And if it is the latter, is the project-based system not at much as fault as the projects themselves?

Evaluation has about it the ring of objectivity, and it seems very petty to criticize an instrument that must, surely, give us a better understanding of what works and what does not. But a closer examination of the process of evaluation suggests that, on many occasions, the instrument is used for quite a different reason. For example, if the project manager feels that certain aspects of the project are going awry, he will tell the midterm evaluation team to pay most attention to the way forward, and to worry less about what has been done in the past. Similarly, a final evaluation can often be used to support the latest fashion in development, for example, to ascribe the difficulties being experienced in project implementation in terms of governance rather than project management.

A detailed study of evaluations undertaken of the Lusaka Squatter Upgrading and Sites and Services Project, which attracted much international attention at the time, shows how the agendas of the authors and/or the agencies commissioning the evaluations have biased the findings. It was clear that, using the same data, it was possible to draw conclusions that looked authoritative but were, in fact, personal interpretations. Thus, commenting on the system of participation in the Lusaka project:

> David Pasteur wrote a study of the project and its organization and management. He was concerned that local leaders might be inclined to steamroller decisions in this supposedly participatory process.[1]
>
> In the World Bank's final evaluation, however, we read that local leaders became less, not more, dictatorial as a result of the participatory process used in the project. . . .[2] (Ledogar) regarded the community participation as having been a success, since it achieved consensus among the many interested groups[3] . . . But the Primary Researcher[4] agreed with Pasteur that the process was not really as participatory as some had made out.[5]

The work referred to lists these different interpretations in some detail, and the factual basis of the data on which interpretations were based. It concludes:

> All these conflicting views are based on a careful observation of the same facts by experienced and honest people whose objective has been to evaluate a process frankly and dispassionately. In doing so, however, they have adopted positions against which the process is evaluated, positions that they themselves may not be fully aware of. There is no stigma against evaluating from a given position: this, after all, is the origin of the classical scientific method of stating a hypothesis and then testing it. Unfortunately for any science that concerns human behavior, however, hypotheses are multi-dimensional and cannot be expressed in such terms that all persons reading them will interpret them in the same way. This is because the issues raised by the evaluations of housing projects are not value-free, and political or human issues that can affect the perspective of the evaluator will presently arise.[6]

But even if there are difficulties in terms of the interpretation of data, and the conscious or unconscious bias of the authors, surely the concept of evaluation is helpful? Does it not show what works and what doesn't?

Well, yes and no.

An unfortunate feature of evaluations is that many are completed after the fact. Those who were implementing the project have packed up and left, so are not there to make use of the findings; much the same is usually true of the project designers. Recalling that projects typically take ten years from design to completion, the policy landscape has changed dramatically by the time any evaluation appears. Even if evaluations are done during the project—for example the typical midterm review—by the time it has appeared, been scrutinized by all affected parties, and the text sanitized to remove criticisms of official action or failure to act, it is out of date.

In Martin's study of the evaluation system used in Lusaka, he describes how painful delays[7] were caused by data processing difficulties and later by the need to have reports "approved" by the concerned agencies before release. Not only did these factors make the findings out of date (as the implementing team had already responded to the same circumstances by changing its procedures), but they also made them likely to be watered down in order to prevent any offence being caused. Thus the very objectivity which is supposed to exist in an external evaluation can be compromised by those who commissioned it.

In a reaction against the sort of judgments just described, there has been an increasing bias toward numerical data. The view is that if human bias affects the objectivity of an evaluation, surely this can be eliminated by the use of scientific method. In such cases, once variables have been established through household surveys and geographical or other data, then the conclusions must be objective. We would urge caution in this matter. Numerical data may give a spurious sense of accuracy and objectivity and may, in practice, hide unwarranted assumptions and sloppy survey methods.

The formula in Box 10.1 is taken from a sophisticated study conducted in India regarding informal settlements. Using socioeconomic data collected from household surveys, this study aimed to determine whether people would prefer to remain in an upgraded area rather than be moved.

Box 10.1　Mathematical formulae as evaluation tools.

---

possible ward choices to ith household. For estimation we will assume that ε_w^i is additively separable from the rest of the utility function, and has a Weibull distribution, which leads to a conditional logit specification,

$$P_w^i = \frac{\exp(\alpha_X^i X_w^i - \alpha_D^i D_w^i - \alpha_p^i p_w + d_w)}{\sum_k \exp(\alpha_X^i X_k^i - \alpha_D^i D_k^i - \alpha_p^i p_k + d_k)}$$

in which P_w^i is the probability that ith household chooses ward w.

We use this illustration to both intimidate any nonmathematical readers and to illustrate the complexity of such calculations. The authors of this study would probably be the first to admit, however, that the weight given to different factors is open to dispute, and that all modeling has the disadvantage of simplifying a very complex situation.

Studies of this nature are, of course, unable to take into account personal factors that affect individual families. Thus if a householder had an elderly mother living nearby, he would probably choose not to move, no matter what the inducements might be; whereas if someone was offered a new, well-serviced site next to his workplace, he might choose to go even if it meant losing a supportive circle of friends and neighbors.

Above all, evaluation is an external activity, and is thus, in most ways, removed from the function of development. To this we add the typical delay in completing evaluations, which further removes it as a tool to be used to make development work.

If we consider evaluation to be overrated as a development tool, how can we review progress, systems, and impact? No one would deny that these are important.

There are few clues in the literature, but our experience provides many useful pointers. The system must have the following characteristics:

1. The beneficiaries must be the primary arbiters as to whether the system is working for them.
2. The feedback must be informed.
3. The feedback must be prompt.
4. The system must be transparent.
5. The process must be supported by all stakeholders.

For many people the involvement of the beneficiaries in this type of work is a troublesome concept. What powers would they have? How would they access information? Could this process not be hijacked by a few leaders and manipulated for their own advantage?

Let us sketch a scenario to examine how these (very legitimate) concerns can be addressed. We assume that it has been decided to upgrade a road within a settlement. The project has gone through two stages, and is in the middle of the third, construction, phase.

Stage 1: The route of the road, the width of the road reserve, and the standard of construction of the road have all been agreed with the full participation of the community. The houses of a few families will have to be demolished to make way for the road, and they have been given plots elsewhere in the settlement. They have fully consented to the move.

Stage 2: The contract for the construction of the road has been awarded by the local government. The community is informed and provided with a program for the work.

Stage 3: The contractor has started work on site. Bulldozers are clearing the route and have already demolished two of the affected houses.

We assume here that:

- Community facilitation activators (BBs) have been working with the community.
- A project monitoring group has been established by the community.
- The project monitoring group has been kept informed throughout the process regarding the design standards, the procedure whereby the contractor was appointed, the rules of the contract, and the technical procedures for supervising construction.
- The monitoring group has communicated to the community that if there are any concerns, they should be informed.

They have four complaints to deal with at the present time:

1. The families whose houses were demolished were told that before they lost their houses they would have time to build new ones. This has not happened. The contractor demolished their houses two weeks early, and also damaged some of their goods.
2. The local party committee has said that the road is too narrow: they insist that it be increased so as to allow buses to use it.
3. The bulldozers have created banks alongside the road which prevent trucks making deliveries to the small shops, and no alternative arrangements have been made.
4. The contractor's employees are arrogant and have been making rude remarks to women passersby.

At this stage, the reader will ask what this has to do with evaluation. Very little, of course: it has to do with project fit. If the project fits well, then people will be satisfied. Our point is that rather than wait for an evaluation to determine whether project objectives have been achieved, the project should be subject to continuous review. If the review mechanisms are properly established, then the likelihood of a good fit is greatly enhanced.

1. Reverting to the list of previous problems, they highlight a number of issues.

    The contractor needs to understand the importance of the issues regarding, for example, demolition of the houses. He should be invited to attend the review committee both to be informed of their concerns and also to be able to explain how he will address the issue in future. This affects the demolition issue, the blocking of existing roads, and the conduct of the workers.

2. The question of one section of the community complaining about a decision after the contract has been awarded needs to be addressed urgently. It should be emphasized that the committee (and the community as a whole) needs to be well informed. On the one hand, they need to appreciate the impact of trying to change standards midway (in terms of additional costs, delays, etc.), and, on the other, they could maybe be educated and shown that buses can use the road, even though it may look too narrow now. Alternatively, a compromise might be agreed to, for example, to widen the road in places to allow wider vehicles to pass.

In a good government system, the role of the project review committee will have been agreed up front, and will have been written into the contract documents. It will not have any powers to instruct the contractor but will work with it. The resident engineer will sit on the committee and will advise on matters of contract, price, and procedure. The BBs will assist and advise throughout and will be responsible for ensuring that the committee's work is discharged effectively.

The difference between this approach and the formal evaluation approach is that one is internal and the other is external. Our exploration into cognitive dissonance showed that external criticism provokes resentment and resistance, whereas self-criticism can quickly lead to change. An old fable presents it differently, making the point that the use of confrontational tactics and the use or abuse of power are not always effective:

The wind and the sun are having an argument about which is stronger. The wind says "I'll prove I am. See the old man down there with a coat? I bet I can get his coat off him quicker than you can."

So the sun went behind a cloud, and the wind blew until it was almost a tornado, but the harder it blew, the tighter the old man clutched his coat to him.

Finally the wind calmed down and gave up, and then the sun came out from behind the cloud and smiled kindly on the old man. Presently, he mopped his brow and pulled off his coat. The sun then told the wind that gentleness and friendliness were always stronger than fury and force.[8]

Self-evaluation and monitoring do not just happen; they must be properly structured and managed. A case study from Nicaragua gives us an insight into a comparatively successful case, although it also provides an example of what can go wrong.[9] The project was a self-help housing project comprising 48 units. The houses were to be built by a collective of the beneficiaries—that is, with their own hands. In a spirit of rejection of the formal structures of evaluation, and of self-reliance, it was decided that the task of evaluation and monitoring was to be undertaken jointly by the beneficiaries and the ministry officials. This participation in the monitoring process "helped to eliminate any feeling of suspicion and mistrust towards the technical staff and made community leaders more enthusiastic about evaluation in general."[10]

However, the first monitoring report identified some role definition issues in the way that the work teams were organized. The men had typically much more experience in building, and tended to take "managerial" roles. The report also criticized some people for not pulling their weight in terms of construction work.

> The contents of this report reflected the opinions of most of the community team leaders but greatly angered many of the rest of the workforce, particularly the women. The solution to the problem proposed by the men was that the women should send men to represent them at the building site. This infuriated the women who had attended the education course, where the emphasis had been on working together.[11]

In the long run, however, these issues sorted themselves out. Different approaches were adopted over time, and different leaders played a role in the monitoring.

> This variation in monitoring approaches might have arisen because the first batch of leaders had been selected more for their knowledge of construction than their qualities as leaders ... At the same time, the individual beneficiaries had also learned that constructive criticism was not to be taken personally but was for the benefit of the program. Above all, leaders won the respect necessary to give them the authority to carry through any proposed changes within the group ... In spite of the difficulties associated with the early monitoring reports, the positive benefits of continued monitoring by the collective were demonstrated by modifications to the collective labor schedule that came about in direct response to the personal situation of the families.[12]

This example demonstrates two important facts. First, where community involvement is concerned there are often problems, and while the aforementioned guidelines we proposed will help to prevent such problems from occurring, the occurrence of difficulties does not mean that the concept will not work. Second, it shows how, in time, the system had very positive benefits which resulted in enhanced performance.

SELF-SURVEYS AS A TOOL IN MONITORING

There is an increasing body of literature about the use of self-surveys, or community-based enumeration, within low-income communities. The following extract illustrates the value of this approach:

It is worth taking a closer look at the community-based enumeration approach, promoted by the international NGO Slum Dwellers International, and practiced in countries such as India, South Africa, Zimbabwe and Kenya, as this has been refined to ensure relatively high levels of accuracy in informal settlement situations . . . Through door-to-door surveys that are conducted by local residents, information is shared about envisaged intervention, and households are able to make individual decisions in relation to these projects . . . The accuracy of the population data is increased through a 'multiple verification' process (repetition of the door-to-door survey).[13]

This enumerative survey is not, in itself, a monitoring tool, though it can be used as an excellent source of base-line data if well designed. What it does do, however, is to include the residents in a system which feeds their personal details, aspirations, and expectations to a higher level, and thereby gives them a perceived and actual stake in the future. As such, it can be used as a vehicle for community monitoring and evaluation during project implementation.

PROJECT IMPLEMENTATION: THE STAFF FACTOR

So far, we have been discussing general matters regarding project implementation and outputs, as it is upon these that evaluation typically focuses. However, there are other matters which anyone who has dealt with the public sector will be only too sensitive to: the conduct of the staff.

One of the reasons that the public sector has earned a bad name is the apparent indifference of the staff to the needs and feelings of the public. A don't-care attitude expresses itself in arrogance, laziness, or even rudeness.

In India, long recognized as having a particularly dominant bureaucracy, the bureaucrat has been characterized as:

a. A perverse God who must be propitiated;
b. A recalcitrant ass that must be driven;
c. A privileged snob, impossible to get the better of;
d. A lazy hound, impossible to bring to book; and
e. (Occasionally) a hard-worked, underpaid and harassed official doing his best under difficult circumstances.[14]

When we think of people we meet at the counter, paying bills or looking for information, these are the bureaucrats whom we think of. But there are other bureaucrats sitting in higher places whose job it is to spend money, take decisions, and manage the public service. The same author has a view about them.

There are four Cardinal Principles of a Decisionarian[15] in governmental administration:

1. If you can avoid a decision, avoid.
2. If you have to, take a decision right away.
3. Once you have taken a decision, camouflage it.
4. If you can, never allow a post-mortem.[16]

How do the beneficiaries, users of the public services, get good service from those that serve them? The participatory monitoring system that we just described can work for that as well. It is remarkable how, when public officials are confronted by their own behavior by the people they are supposed to serve, they not only become more conscientious and hard working, but also enjoy their work more. This is because they feel a relationship, and bond, one might say, with the community and want to help make things better for them.

How can this be achieved? As we suggested, one of the steps is for the beneficiaries to give feedback on the service they receive. For example, that the office was opened late; that such and such cashier didn't give a receipt, or that such a person was favoring certain members of the community. This message would be given verbally to a member of the project monitoring committee, who would then be in a position to bring this to the attention of the person concerned. Another technique is to have exit polls, as it were, to canvass the views of members of the public who have just been served by a public official. An extremely effective way is for a video to be shot, as if on a TV news broadcast, in which people express their opinion on the service they receive.

The feedback does not necessarily have to relate to the conduct of individual members of staff, but can, for example, feature complaints about opening hours, distance from someone's home or work to a pay point, and so on. When such a video is shown to the management, it can help them appreciate the difficulties that their "customers" face, and thereafter to find ways of improving their service.

The style in which this is done is more important than the content. The staff who are working within the community must be included as stakeholders in the monitoring system. From the outset their commitment would be obtained to working with the community to achieve the best standard of service, and listening to the feedback that is received. By so doing they will stand to earn accolades as well as brickbats.

Corruption is another problem which can be tackled with community support: indeed, if the public have an understanding of what is supposed to be done, then they are in a strong position to identify corrupt practices.

Corruption can, of course, take many forms. What we are describing here will be no substitute for the professional audit which might reveal that a senior bureaucrat has taken a bribe to award a contract, or suspect practices in terms of staff appointments and the like.

Box 10.2   The result of inadequate monitoring and follow-up.

The site and services project in Nairobi described in Chapter 4 is pertinent here. Peter is one of the original beneficiaries of the scheme. He alleges that there was corruption involved in the allocation of construction loans for this sites and services scheme:

"Everyone wanted a loan, and there was enough money for every beneficiary to get one. But not everyone got it. So where did the rest of the money go?

"It was a cash transaction, and for every loan that was issued, the loan officer pocketed KSh5,000 ($700). I was made to sign for KSh36,000 ($5,000), but was given only KSh31,000. If the money got over, and you were at the end of the line, you received nothing at all. One day, a woman made a noise about signing for more than she had received. The police came, and found stacks of cash—KSh180,000—under the chair of the loan officer. From then on, cash transactions were stopped and checks began to be issued. But this new system posed another problem: most households got checks that bounced, and had to wait long periods for the loan to be processed, if at all. Many didn't get anything in the end.

". . . . According to the initial terms of the loan and plot allocation, I was to receive title to the land after 25 years. Per the original terms, my payments finished in 2001. When I went to the City Council to inquire about the title deed, I was asked to bring KSh7000 more for a processing fee. I have been making that payment incrementally since, but now there is a rumor that, as of January 2005, the Council has added an extra clause, demanding an additional KSh50,000 for award of the title. We can't increase rents to compensate for this extra cost either; this is because the demand for rental units is reducing due to security problems here, and also because there is a lot of cheaper rental housing available in the poor quality multistory buildings. So how are we going to raise so much additional money to pay for the title deed?

". . . . Also, the process is long: the first step is to get verification of familial relationship from the community elders. The application then goes to the Chief's office for endorsement. From there, it goes to the Nairobi City Council for final approval. The process can take up to two years, and many bribes along the way."

". . . . Although most of the original owners still retain their plots, some original beneficiaries sold them to commercial interests. It is these plots that typically have the multi-story structures on them, with absentee landlords. Some of these new owners have also built on public spaces, so the children have no playgrounds left. Due to the high density in these buildings, they use up more water, constraining the overall supply in the neighborhood."

Box 10.2   (continued)

"... Things might have been better if there had been better controls in place, and more interaction with the beneficiary community, for example, a project office where the residents could go and express their concerns. ..."

Source: Ashna Mathema, *Nairobi's Informal Settlements: A View from the Inside*, Background Paper, 2005 [Unpublished. Funded by the Norwegian Trust Fund/ The World Bank]. The name has been changed.

But it can make a difference in terms of, for example, allocation of plots for people within the community, the implementation of the contract in terms of the accuracy of payment certificates, the quality of work done, and so on. Within the public sector it can help reveal misuse of public-sector vehicles, theft of stationery, and so on.

It can also, for example, have an impact on corruption in the planning and house-building field, where unscrupulous landlords, for example, pay off community leaders and/or building inspectors to allow building which encroaches on others' land, or is unsafe, or in other ways infringes on the rights of others.

Once more we revert to the question of transparency. If the community, and particularly the monitoring committee, understand what should be done, they can be a very effective power in ensuring that it *is* done. Once more, however, we must state that there are risks. It will be crucial to avoid setting up a confrontational position with the contractor which can lead to accusations of breach of contract due to interference and so on.

The parties must show each other due respect. The role of the resident engineer, who is supposed to check the quality of the work, and approve payments, must be understood. But he must also understand that the community has a right to be satisfied that the quality of the work is in accordance with the contract. Speaking as professionals, we know that this is not a comfortable arrangement, as we like to feel that we should not be subject to such scrutiny. But in the long run it has substantial benefits, not least in terms of limiting the scope for corruption. Box 10.2 illustrates this with an example of a sites and services project in Nairobi.

THE INDONESIAN EXPERIMENT

The most important work regarding the use of community monitoring to reduce corruption was conducted in Indonesia in 2004–05. The objective of the work was to test the hypothesis that increased transparency acts as an inhibitor of corruption, and to compare it to the impact of threats of audits and criminal prosecution. The projects being studied were ones in which the community was given funds to build roads. Community management committees took responsibility for buying materials, hiring labor, and managing the works.

The study took place in 608 villages. Some villages were told that the project would be audited by government auditors, and the results of the audit would be read to a village meeting when they were completed. The prevailing values in the villages were such that anyone caught engaging in corruption would suffer substantial social sanctions.

In a second category, the villages were all invited to accountability meetings at which they were invited to participate directly in monitoring the project, and thus expose the community managers of the projects to increased scrutiny.

In the third category, the villagers were given anonymous comment forms, which were placed in sealed boxes and publicly opened at accountability meetings.

The problem with any survey regarding corruption is that a major part of the energies of its perpetrators will be devoted to preventing it being uncovered. On the assumption, therefore, that corruption diverts funds from the activity of the project (in this case road building) to personal consumption, the work measured the value of roads as built and compared it to the cost. In those cases where the standard of road constructed was lower than the cost would suggest, they were presumed to be affected by corruption.

The results are very interesting: the expectation of an audit was the most effective deterrent, and were particularly effective where the village head was up for reelection. In the "invitation villages" there was much more open discussion about corruption, while in the "comment-form villages" the impact was less.

Interestingly enough, the impact of the two participatory methods on reducing corruption in materials purchase was slight, but there was a substantial reduction of missing labor expenditure. The author concludes:

> There are two potential explanations for why increasing grass-roots participation would have impacted labor expenditures but not materials—either community members had better information about labor expenditures than about materials, or community members had a greater incentive to monitor wage payments than materials payments (since they were the ones working on the project and would be the beneficiaries if wage payments had been stolen). In fact, the results suggest

that villagers had less information about missing labor expenditures than about missing materials expenditure. Furthermore, the invitations treatment was most effective when the workers came from inside the village and participated in monitoring. Combined this suggests that grass-roots monitoring may be most effective when a subset of people personally stand to gain from reducing corruption, and therefore have strong incentives to monitor actively.[17]

The importance of this work is that we cannot expect community participation in monitoring and evaluation to be the complete answer to corruption. How it is structured, how information is communicated, and the transparency of the system as a whole are all very important factors.

Increasing transparency is clearly the main challenge, but this should be combined with external scrutiny whether by auditors or, as the following case shows, the media. The ministry of finance in Uganda was facing problems in relation to its disbursements to primary schools. Excluding teachers' salaries, a tracking survey revealed that only about 20 percent of the funds disbursed were actually reaching the schools. Audits were not proving particularly effective, so the permanent secretary tried a totally different approach—bottom-up scrutiny. Each time the ministry of finance released money it informed the local media and also sent a poster to the school setting out what it should be getting. When the tracking survey was repeated three years later, it revealed that 90 percent of the money was now getting through. Public awareness was a very important factor, but the local media played a crucial role in bringing about this awareness.[18]

If there is a message in both these case studies it is that monitoring can be embedded within the community, and as such it is a much more powerful tool than the typical expert-driven process. But for its full potential to be realized it is necessary to locate it within a broader framework in which there is transparency of information and efficient communication.

IMPACT EVALUATION

Another approach to evaluation is to concentrate on the *impact* that a project has had, rather than the degree to which project objectives were met. In the best of all worlds these might be the same, but in practice project objectives are often not well defined, or inappropriate ones are used. The project might be based on certain procedures or management systems which do not work in practice. It would be strange, but not unheard of, for an evaluation to criticize the project for deviating from such a flawed implementation plan even though by so doing, the higher level project objectives were met.

Impact evaluation therefore concentrates on the degree to which the project has had an impact—or, to use nonjargon, "made a difference." The problem of how to measure the impact nevertheless remains, and it is always easier to look for numerical indicators than to evaluate the fuzzy components which might be more important to the beneficiaries.

As a response to this, impact evaluation makes use of community panels which allow the community to establish its own criteria for evaluation and monitoring. These panels then determine the extent to which those criteria are being met. This method offers scope for unfair criticism of the project staff, but if established within a context in which the participants see the evaluation and monitoring as a component of their own management of the project process, it can be a very powerful tool.

Such qualitative feedback from the community can be the most informative and cost-effective first step in identifying when things are going awry.

11 New Ways of Working

This chapter draws together the threads of the book as a whole, which has shown that real development requires a new way of working. It suggests that sustainable development requires continuity, certainty, and trust between all the parties involved and shows how this can be achieved. It also shows that the power relationship between the parties must be changed so that people can have responsibility for and control over their affairs. It examines the potential for creative engagement by the private sector and describes financing mechanisms which can be used to make such systems work.

This chapter starts by looking at the concept of governance and especially the concept of good government. This leads us to an analysis of what roles all the stakeholders in government can play to the best effect. In this we include the state, the community, NGOs, and the private sector. After defining these roles, we can revisit the role of money in development—and possibly come to a surprise conclusion about its importance and impact. We try to anticipate what problems are likely to occur, and suggest ways of tackling them. Finally, while we are confident that the principles we discuss have universal applicability, we warn against prescriptive solutions: specific solutions must be developed within the society in which they will be used.

DEFINING THE WAY FORWARD

What then is our prescription? It is that we must start and finish with the needs of the people, and the better we can fit our services to them, the more effective development will be. This message may seem disarmingly simple, and to many it will sound easy. This is far from true: it is a difficult message, and it will only sound easy to someone who has not tried it. But to those who have tried it and found it difficult, we hope that the book will help them understand where problems might have developed, and to provide a few tools which will help them in the future.

ROLE OF THE STATE

The state has not proved itself particularly judicious in the use of its powers, and the hardest thing it has to do is to restrain itself from abuse of its own power. The most typical form of abuse in terms of informal settlements is to demolish them, but there are many other less obvious ways in which powers can be misused.

What then is the role of the state[1], and how does this fit into a system in which the beneficiaries of development can have the freedom to decide how and what development they want?

Chapter 6 described the value of partnership as a participatory concept, and as a way of making things happen, but it did not delve in any depth into the role of the parties in such arrangements. Chapter 7 looked at the role that government can play in facilitating development. Chapters 8 and 9 looked at the how to oil the wheels of collaboration between community and state and support community-based development. But the most difficult aspect, and the one which can, and does, cause the most difficulty has not yet been touched on: the power relationships between state and community.

There is no doubt that the state has a valuable role to play. It raises money through taxation and spends some of that money on "development". States, like private-sector companies, have rules about how their funds can be used, and these rules are typically designed around conventional procurement procedures. These procedures are unlikely to be appropriate for participatory projects both because of the typical silo-type budgeting system (which limits the use of funds designated for one sector to be used on another one) and also in terms of the management systems which conventional project implementation requires. There are thus two aspects which must be examined in some detail:

- How can state funds be applied most effectively in community-based development?
- What model is most appropriate to account for the funds so used?

What is required of the state in development? We have already drawn attention to the needs of communities in terms of their lack of capacity and their rights. Are there certain types of development which communities cannot, or are unsuited to, undertake, and which the state can do? Clearly, major infrastructure is an example of this. In particular, the state has a responsibility for the provision of trunk infrastructure and services within the natural monopolies of water, sewerage, and electricity. There may be other areas in which the community feels it has neither the power nor the skills to be effective, and therefore appeals to the state for support.

If the state (either centrally, or more normally, locally) has a duty in respect of these services, it must determine what the people want and how.

With this understanding it can deliver those services in a manner and at the time required. If this sounds like a broken record, that is only because it is so important and must be emphasized throughout at all times. We are *not* saying that the state *must* provide water, sewers, and roads, but that the state must make its skills and resources available to the community to meet the effective demand for them.

This message may seem to be in conflict with the rights-based approach to development which defines a minimum standard to which all people are entitled.[2] Implicit in this entitlement is the assumption that the state should therefore go ahead and make the provision. This, we consider, can be an inappropriate message, in that it places the responsibility for provision solely on the state, regardless of the wishes of the recipients. Such issues cannot be divorced from financial capacity: in many cases the supply of services brings costs, and far too often the sweetness of the service contains the bitter pill of costs which people are ill-able to afford. This has been the downfall of many projects, especially those which seek to move people from what are labeled slums to new, fully serviced sites.

To return to our theme: let us examine a process in which the community requests the state to assist them in improving their settlement. In order to meet this role, local government must have a means of communicating with the community, and must open channels so that the community can communicate with it in a spirit of frank and open dialogue.

Thus local government transforms itself from the traditional *government* role, which implies the making and enforcement of rules, as well as the obligation to maintain services to its citizens within its community, into a *developmental* role. Developmental, in this context, has a very specific meaning: the obligation to assist and support all communities within its jurisdiction to enhance their welfare. Within the context of this book, this means the infrastructural services and (possibly) housing that are typically lacking in informal settlements. But the broader context demands much more: a developmental local government looks at the needs of the informal sector in terms of business, skills, support, and linkages, as well as the access of disadvantaged communities to education and health services, for example.

Consider, for instance, the case of South Africa, where the term *developmental* was heralded by the new government to suggest such a new relationship with communities. Most of the correct ingredients are present in the legislation to permit it to develop as such. But the essence of the system, certainly as defined here, has been lacking. Not only has there been a very poor record in terms of allowing communities to participate in decision making—let alone project design—but the record in terms of public interest communication has been very bad. As a result, there have been frequent mass demonstrations, many of which have been due to a simple lack of information about what the municipal council is doing. This is because no one has either the responsibility to involve the communities in local governance and development or the skills to do so. Politicians typically claim this territory for themselves, but they often project themselves as, for want of a better expression, "an arm of the law"; in other words, they largely satisfy themselves with communicating decisions made by the council to the electorate. Their other role, that of complaints hotlines—"Councilor, the road in front of my house has got a pothole", "Councilor, my water bill is wrong"—can undermine the workings of a proper system unless it is used as a last resort.

The relationship we propose is one in which the resources of local government are put at the disposal of the communities. In order to do so, local government provides staff who are professionally trained to act as neutral intermediaries. Their job is to, on the one part, foster and support participation and community action, and on the other, to ensure that the services and skills of the local government machinery are used to the best effect in the service of that community. Thus there is a complete *volte-face* by government from control to support.

Development agencies often equate spending money with development. It is hoped that readers of this book by now will agree that this assumption is far from correct. Expenditure of money can, at times, *retard* development, and the way in which money is spent can cause conflict and delays. At the macro level there are plentiful examples of development aid being misappropriated. At the local level the situation is less clear, but money-led, as opposed to needs-led, development is very likely to be disastrous. Far too often, for example, when there is a health scare, action is taken in respect of informal settlements which is retrogressive and causes damage to the interests of the individual households affected as well as to society and the economy as a whole, such as the demolition of the settlement and the removal of its inhabitants to an out-of-town remote location "in the interests of health."

In describing the many forms which participation can take, much was made of the value of the term *partnership*, and it is to this that we now return. In brief it is that the state must transform itself from being the controlling instrument which it was initially designed to be into one which embraces the need for those it serves to be considered not as "subjects"[3] but as collaborators and partners.

The partnership concept has now entered into the lexicon of development under the moniker of public-private partnerships (PPPs), but although these may have some relevance they do not provide the full flavor of what is under discussion. PPPs are essentially civil contracts (so far so good) between the state and private-sector companies. In some cases the private sector provides capital, in some cases it provides skills, and in some cases it provides both. The contract will typically include performance indicators against which the actual implementation of the contract can be measured, and the contract may be subject to verification by impartial experts.

The PPP concept of the two parties entering into a contract for the performance of specific tasks is, however, not always a partnership. Often it consists of the state *using* the private sector in order to perform functions which the state has an obligation to do. In such cases the initiative lies with the state regarding what form of assistance by the private sector to use, and when and how to engage with the private sector. The term *partnership* is therefore something of a euphemism for a convenient delegation of powers and duties, in return for which the private sector receives payment.[4]

The concept being proposed here, however, is different from the PPP model. Our premise is that what is necessary is an arrangement whereby the state does not have the discretion as to whether to enter into the partnership, but a duty: that is, there is a constitutional duty imposed on the state to behave in a certain way.

Examples of the state being constitutionally required to give money for certain purposes can be taken from Ghana and South Africa. In the former, the government is constitutionally required to devote 5 percent of its annual budget to the District Assemblies Common Fund.[5] The individual district assemblies each receive a quota based on a formula, which includes criteria such as population density and incomes. Thus each local government unit has the *right* to a certain level of grant funding every year. The South African system is very similar, though the constitution shrinks from specifying any fixed amount or proportion. Instead it refers to an "equitable share of revenues raised nationally."[6] As in Ghana, how the sum is distributed is based on a formula, while the actual amount is decided by parliament in an annual Division of Revenue Act.

The importance of these formula-based provisions is that they remove the discretion of the politicians and civil servants alike to favor (or discriminate against) individual local government units, on the basis of their political allegiance, for example.

Similarly, the system being proposed here would remove the option by the civil servants as to whether to enter into the partnership or not. In fact, the shoe would be on the other foot: the community would now have the right of access to funds and the discretion regarding how they will be spent. The partnership would therefore lie in terms of the state being a funding partner, and the community being the designer and manager of the work.

Traditionalists will throw their hands in the air at such an extreme reversal of roles, but the case studies illustrate how this can work. What follows is therefore a proposal for new ways of working which are not money-led, nor project-led, but are sustainable in the very best sense of the word. This way of working gives the residents of informal settlements a meaningful role in devising their own answers to their problems, and gives them a political and economic stake within the system of governance. To do so requires local government to treat them as partners, an act of trust that is not always easy yet absolutely necessary to make a genuine difference. Moreover, when we talk about trust we do not just mean trust in the social sense, but the financial sense as well: the community must be given control over resources.

No public agency ever has enough money to serve everyone's needs. Capital for development must therefore be rationed, and if costs are recovered at all, it would either be through user charges or formal loan mechanisms. The needy and disadvantaged must compete for resources against other claims: the major highway, the new market, the sewage works versus street lighting in upper-income areas. But it must be assumed that they will be given a fair slice of the cake, especially now that the Millennium Development Goals are receiving so much attention. The traditional budget system is based on projects: so much for the new highway, to be spread over three years; so much for the market, and so on. Allocating the money is one thing, but how is responsibility for its expenditure to be managed?

Before attempting to answer this question, the principles by which it would be distributed must be considered. For example:

- Predictability: although it is not necessary for allocations to be made annually (indeed, this might not be helpful, as the sums could be too small to spend effectively), it should accrue on a regular basis so that, for example, there are payments on a rotating three-year basis.
- Amounts would be determined on the basis of the population and the need, using an objectively verifiable formula.[7]
- The money would be granted to a community trust from which is would be disbursed directly to suppliers and contractors. In this way no funds would be handled physically by the community members.
- Payments would be released in stages by local government on presentation of satisfactory documentation.

Clearly there are many difficulties in such an arrangement, and there will be temptations to corrupt practices. Typical ones are to inflate contract prices and receive kickbacks, or accept bribes for the award of tenders. Unfortunately, these practices are found at all levels of government, and even in the private sector. Community-based money management should not be expected to be particularly different.

However, experience has shown that the following principles are very important:

- Wherever possible the community must have a financial and/or physical stake in the work. Thus, a condition for accessing the funding could be that the community must either provide, say, 10 percent of the grant sum up front, or must provide labor in lieu of it.
- Wherever possible money matters should be managed by women. They show themselves to be better custodians of resources and generally more trustworthy. The experience of microlenders with a huge variety of people of different cultures and at different economic levels has demonstrated this in many ways: "Women borrowers proved to

be more disciplined and resourceful—their payments came in more regularly."[8]

- It will usually be necessary for the group to have gender balance, but women's effective participation should be actively supported. One way of doing this might be to place responsibility for the account in the hands of those who have already proved their competence in money matters, for example, in a group lending scheme.
- Temptations should be removed: for example, there should be no cash.
- Deterrents should be in place: for example, all the committee members should have a very clear idea of how auditing works and the chances of them getting caught out if they embezzle or enter corrupt relationships.
- Transparency is essential, so as to limit the potential for backroom deals.

If such safeguards can be established, how would the system work? Let us assume that a road is being upgraded. The location and design of the road have been agreed, and it is known to be within the budgetary allocation which is due to be released within a few months. Engineers appointed by the local government, with support and consent from the community, have completed the working drawings and tenders have been received. The community follows public-sector bidding guidelines in how tenders are awarded, and advised by the consulting engineers recommend a contractor for appointment.

At this stage the involvement of the community has been to decide, with technical advice, the route of the road and the standard of construction. Now, tenders are received and it is found that prices are above the budget. There are now three choices, and they have to be made within, say, 90 days if the tender price is not to increase.

1. Reduce the length of road to be built, and include the remainder in the allocation to be received within three years.
2. Reduce the standard of the road, for example, by making it slightly narrower, or with a lower standard of surfacing.
3. The community makes up the shortfall.

From this short scenario, readers will get an idea of the enormous value of placing the community in charge, in terms of empowerment and the taking of responsible decisions within the framework of real financial constraints. Engineers may, as was suggested in an earlier chapter, find the risks of such a delegation of power quite frightening. In prospect could be delays in decision making which jeopardize the contract award, scaring off contractors due to the unfamiliar nature of the arrangement, and so on.

The opposite is likely to be true. People will not delay a decision when it affects their immediate interests—this sort of behavior is much more likely when a body like the local council has to take a decision for a community of which no councilor is a resident. And as for contractors' fear that community involvement in the contract will lead to disruption and irresponsible interference, experience shows that such contracts are far more likely to run into problems where the people are *not* involved. For example, in Lusaka, the contractors welcomed the communities' participation in the process, and there were no conflicts.

To continue the scenario: the contractor is appointed on the basis that the standards will remain the same and a shorter length of road will be built. Thereafter the contract is supervised by the resident engineer. A community member volunteers to work with the resident engineer to see how quality checks are undertaken, and how work is measured for payment. An officially designated community person sits as an observer in the official contract site meetings, and the engineer reports on a regular basis to the project monitoring committee.

The contractor submits a payment certificate to the project monitoring committee, which then approves it and passes it to the community trust fund for payment. The trust would follow similar financial management guidelines to those applicable to any fiscally prudent organization, whether public or private. For example, the check would be made out by a private sector bookkeeper appointed for the purpose, and would be signed by the treasurer and chairman of the trust. A complete paper trail would be available for audit and public scrutiny.

ROLE OF NGOs

NGOs can often play an important role in such arrangements. Many have the skills to facilitate community involvement, to help people manage money, to mobilize community self-help, to train the leadership, and the like. But to be effective their role must be clearly defined and understood.

The gap in governance caused by the Reagan-Thatcher antigovernment movement, which is now given the label of *neoliberal*, was such that many essential functions in society fell through the gaps, especially those which affected the poor and disadvantaged, and most especially the urban poor.

Into this gap came NGOs, which had two very important qualities. For the beneficiaries they represented support and advice, and for the governments they removed an uncomfortable responsibility from their shoulders. They could also be used to play a management role. An example of this is in Santiago de Cali, Colombia, under the AOISPEM program (Self-Construction of Public Service Infrastructure Works with Delivery of Materials). In this project a private nonprofit foundation provided management skills, community contacts, and credibility. A municipal public-works company provided technical skills and finance; the community provided labor and monitoring; and building material companies provided materials at reduced cost. This arrangement allowed the costs to be kept at about 60 percent of a conventional contractor-built project.[9]

More recently, the large international NGOs such as CARE, Oxfam, Save the Children, and World Vision have been acting as channels for bilateral foreign aid, and as such are, in most senses of the word, developers, not facilitators. This has changed their role completely: they come as a package, and though it may generally be a good package, it nevertheless includes constraints imposed by their funders. This will normally limit the flexibility with which they can respond to the needs of the communities they serve, and may, indeed, put them into conflict with the community.

By crossing the line between servant and master, the role of NGOs, in the views of many, has been compromised. Even at a local level NGOs have increasingly been co-opted by governments to act as project managers because they offer a relatively reliable source of manpower and fiscal prudence.

One commentator has taken a particularly critical view of this development:

> Major NGOs (are) captive to the agenda of the international donors, and grassroots groups (are) similarly dependent upon the international NGOs. For all the glowing rhetoric about democratization, self help, social capital, and the strengthening of civil society, the actual power relations in this new NGO universe resemble nothing so much as traditional clientelism. Moreover . . . Third World NGOs have proven brilliant at co-opting local leadership as well as hegemonizing the social space traditionally occupied by the left.[10]

This move into commercialism has been prompted by the reluctance of charities to fund NGOs' management expenses on a routine basis. They now like to see NGOs run like private-sector organizations which tender for work on a competitive basis and thereby reduce the burden of running costs. This emphasis on accountability and performance is in conflict with the concept of a community service organization with typically nonquantifiable objectives.

The World Bank has also expressed doubts about the role of NGOs.

> Recognizing that NGOs, like markets, can be important does not mean that they provide the missing link in effective housing policy. Such local systems cannot be expected to replace the systemic sorts of policies needed to assure that many basic services, such as electricity and water are provided . . . Nevertheless, an optimistic message has emerged . . . these groups are showing how participation in housing programs can improve not only their immediate housing conditions,

but also contribute to the creation of, in the words of a recent study, the "capacity to aspire."[11]

We are neutral on NGOs in general: on the one hand, we are aware of the skepticism with which many communities view them, while on the other we know that many very experienced and talented people work for them. However, we do not have any reservation in suggesting that NGOs can be used as an initial reservoir for skills regarding participation which can be used until properly trained local government manpower is available. If they are to be used as that, the staff would be seconded to local government, and would not be seen to be working for the NGO itself. This would be the starting point for good government within the communities.

There are other important roles which NGOs can play. One of them is to play an advocacy role and support the needs of the disadvantaged; for example, orphans, exploited tenants, or small-scale traders whose livelihood is threatened by zealous law enforcement, and so on.

Another very important role is to undertake and/or support training activities. They—especially some of the larger ones—will be able to access the very special skills required for the effective training of community participation activators, and similar work, for example, community builders.

Lastly, international NGOs have made a very important impact in the exchange of information and experience. Such networks have been effective in the transfer of knowledge not only to NGO staff and community leaders but also to public officials. For example, railway officials in Kenya were taken to Mumbai to meet their counterparts.[12] This helped them gain a very important understanding of the value of dialogue with the squatters who were encroaching on the railway lines. Thereafter they adopted the same method in Kibera, Nairobi, and broke a stalemate which had been threatening the lives and the property of squatters and railway users alike.

ROLE OF COMMUNITY-BASED ORGANIZATIONS

CBOs may often represent important interest groups and/or provide a ready-made source of leadership. Of those we have worked with, about half can be characterized by selfless people who will sacrifice their own time—and often money—to help the weaker members of their community. Sometimes CBOs can legitimately claim to represent the whole community and might be effective partners in any development project. The other half, however, have been characterized by self-serving behavior of a few individuals who are exploiting the needs of others for personal gain, especially where the scent of international money is concerned. Hence, it is wise to take any claims to represent the views of the community with cautious skepticism.

Box 11.1   A CBO leader in action: Messiah or mercenary?

A squatter community in Accra had been under threat for years. Various attempts had been made to mobilise the community, and to help them raise funds to improve their conditions and/or fight for recognition. A CBO attracted the attention of several international agencies who were impressed by the oratory and passion of its leader. Although he was not a resident he had managed to obtain not only the trust of the community to negotiate on their behalf with the authorities regarding the future of the settlement, but also to collect their funds for future improvements.

Our household surveys revealed a very different picture. Only a few people had heard of him, and those who had participated in the savings scheme complained that collections had stopped over two years previously and they had no information about the funds which they had contributed. One fact was clear from their responses: they did not see him as their leader, and did not trust the savings scheme.

The role of CBOs as stakeholders, however, must be respected. Special interest groups such as religious organizations, women's groups, traders, youth groups, and the like will have an important contribution to make, and ways must be found of including them in community leadership and decision making.

INGREDIENTS OF SUCCESS

Experience has shown that certain factors can facilitate development. Nabeel Hamdi identified flexibility and spontaneity as crucial components of success:

Spontaneity is vital because most problems and opportunities appear and change in fairly random fashion and need to be dealt with or taken advantage of accordingly . . . When you have run out of resources but not out of problems, you improvise—inventing rules, tasks and techniques as you proceed . . . The third idea is incrementalism. Most settlements grow, consolidate, change, and even disappear in a series of increments.[13]

In addition to these somewhat fuzzy aspects, there are more tangible factors:

Small Scale

The larger the community, the greater the risks of conflict and discord, and the harder is the role of leadership. Numbers are very relative: in some cultures 2,000 households might be ideal, whereas in others a figure five times that might be workable.

Whatever the figure, it is important to deal with a unit that defines itself as such, and this is something that must be determined within each specific context. Often, the community unit is defined geographically, but this can be deceiving, especially where geographical nomenclature is derived from outside the community. For example, the city planners may define a group of squatters as "the Noku River settlement" when referring to settlements on both sides of a river. This description could mask the fact that the residents on one side of the river have been there for forty years, whereas those on the opposite bank are newly arrived squatters, and in fact the long-term residents have a completely different name for their settlement.

So when defining the scale, the concept is more about serving communities which define themselves as such, and less about absolute numbers. More than anything else, small scale refers to homogeneity of purpose and needs, while still retaining sufficient mass, in terms of numbers, to be a viable unit for the purposes of physical development.

Continuity

Continuity is an essential component of development. A possible system whereby there is a trickle of funds annually was described earlier; this would be disbursed every three (or maybe five) years to community-defined projects. The important part about continuity is that it allows and supports planning for the future in a developmental way. This gives people faith and allows planning for the future with confidence. It mirrors the way in which normal household budgets are managed: we start, for example, with a one-room house and plan to build extra rooms every two years until a four-room house has been achieved. Thereafter the plan may be to improve in-house services, for example, upgrade the bathroom. Continuity also builds a mutually reinforcing relationship between local government and the community.

From the point of view of development, one of the most valuable aspects of continuity is the opportunity for learning by doing. As described in Chapter 10, community-based monitoring allows continuous reform and evolution of systems by which development is managed. Where there is a practice of steady development whereby as soon as one project is finished, planning can start for the next one, then learning by doing comes into its own. Together with this there is an increased sense of responsibility for the product and maintenance.

Partnership

As has been repeatedly emphasized, an emphasis on community participation does not mean that communities have to be self-reliant and autonomous.

They need and should be supplied with the same services as all other citizens, for example, schools, clinics, police services, and so on.

Similar principles apply to these services as to infrastructure. In brief, it is as harmful for state agencies to behave as if they have no need of involvement by the community as it is for the community to say that "providing those services is *their* problem." An easy example is that of policing. A community that offers no support to the police cannot expect them to do as well in controlling crime as one which does support them.[14] The concept of the community policing forum has been shown to be effective in getting the public on the side of the police in some South African neighborhoods.

Sharing Responsibility in Money Matters

The value of community contributions in cash or in kind was previously described. This creates a sense of responsibility and ownership in regard to the development taking place. The point does not need to be labored but it must be included as an important component of community-led development.

Communication is Key

As we showed in Chapter 6, no matter how many public meetings are held, and how many leaflets are printed, if communication is poor, then things will go wrong. Good communication ensures that people know what is happening and what commitments have been entered into on their behalf. They know what to expect, and what their leaders are doing. This, in turn, means that leaders are less able and/or less tempted to get involved in any corrupt practices.

Chapter 6 referred to the techniques of mass communication, in connection with community consultation. These are useful and should not be overlooked. But in connection with communication *with the poor*, different problems emerge. For example, they may be:

- Too busy trying to survive to attend public meetings;
- Be illiterate;
- Be very suspicious of statements by the public sector due to past experiences;
- Be too shy to ask questions in public.

To deal with this problem, communication must be on a face-to-face basis. This can be done effectively by electing a leader from each small group of up to twenty-five households, who will be trained to communicate information downward and upward.[15] The *downward* communication is to inform the community what is happening, and the *upward* communication is to give feedback to higher-up leaders regarding what the residents of that group are thinking, and how they respond to the opportunities or challenges of any proposed development. Thus a sort of pyramid scheme, on the lines of a Tupperware sales system, is established, which allows rapid and effective communication to take place.

In the context of the problems of the poor, this system works extremely well. Because the communicator is known to the people, he or she is addressing the question of shyness, which is greatly reduced. But for those who are still too shy to voice their concerns or opinions, they can be seen privately. Also, because the person communicating is from the ranks of the group themselves, the problem of distrust is greatly reduced.

The question of needing to devote all available time to earning money is also greatly reduced because the meetings will usually be held in the evenings, within the neighborhood occupied by people concerned. They can therefore continue to look after babies, and even prepare food, while the meeting is taking place.

Clearly, illiteracy is no longer a problem here, because communication is verbal (though backup written information is highly desirable).

Right Staff and Right Relationships with Local Government

While some readers might have agreed in principle with the precepts described in Chapter 7, they may have had major reservations about whether it can work in practice. In particular, the concept of local government being servant and not master might arouse considerable skepticism. There are examples, some of which were discussed in earlier chapters, which demonstrate the viability and sustainability of the principles advocated here, either in part or as a whole.[16]

However, to many readers, the principles described earlier might sound either unrealistic or difficult or both. Because a new approach is required to management within this context, and because this is a contentious issue, it requires much fuller treatment.

A NEW MANAGEMENT SYSTEM

To be effective, management of participatory projects must be able to respond to the communities in a responsive way. This requires a rethink of conventional management structures. Let us start by stating the principles:

- The clients are the residents.
- The duty of local government is to provide services such as infrastructure.
- Whatever needs to be done can best be articulated by the residents, as advised by the staff of the local government.

- For this to work properly there must be staff whose job it is to act as an intermediary with the community.
- These staff, working with the local government technical staff, are in the best position to recommend what can be done.

This inverts the normal decision-making structure, where the person on top has the last word and whose decision is passed down to those below, in a system that bureaucrats understand and are expert at manipulating. There is a good military analogy to illustrate the point. In military parlance, the overall strategy is determined by the generals, but the tactics are left to the frontline troops. This is for the obvious reason that there is no time in battle to refer a decision back to the headquarters. In development the details of the situation are different, but the effect is the same: the overall project objective has been determined, but the means of achieving it must be flexible.

The two systems can be contrasted. The top-down organogram is shown in Figure 11.1. Thus, as all good bureaucrats will know, a draftsman cannot take a decision regarding the design of roads: his drawings have to be approved by the assistant engineer, and then the design engineer, and finally by the chief engineer.

Figure 11.1   Top-down organogram.

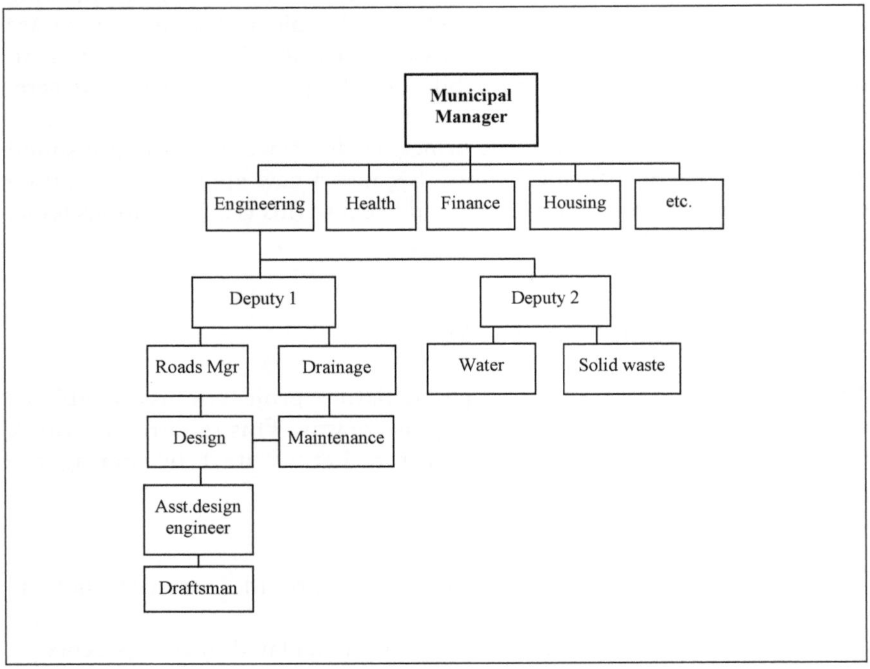

Figure 11.2 Bottom-up organogram.

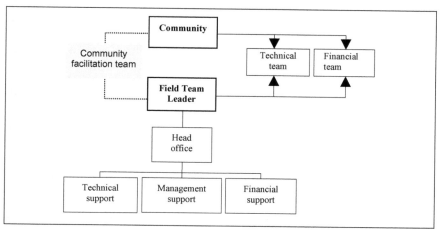

In Lusaka a different approach was taken: the field team leader (FTL) was the one who ran the operations in the field (see Figure 11.2). He was the focal point at which all interest groups in the field converged, including the engineering contractor, the community development staff, the community and individual households, the accounts staff, and building materials stores operators.[17]

Let us say that the contractor comes to the FTL with a grievance: a householder who had agreed to move has changed her mind, and the contract is being delayed. The FTL then informs the community leaders, and together they visit the offending householder. They find that she is very old and infirm, and simply cannot manage the stress and physical hardship of resettlement. It is not a question of money, but of physical strength. The leader, with the FTL, then calls a meeting of all the families in the vicinity to get there ideas about how to resolve the issue. Two options emerge: either the community helps her to move, or the route is deviated. After much pleading by the affected person, it is agreed that the route will be changed.

The FTL is therefore mandated to go to the head office and ask the engineer to change the drawing and issue to necessary instruction to the contractor.

In this small scenario there are three points which are different from the norm. The first is that the process of solving the problem is taken by the community in the knowledge that they have the power to do so. But if, for the sake of argument, the engineer in the head office informs them that changing the route has other implications, then he will refer the decision back to them. The FTL will, in these circumstances, insist that the engineer attend the meeting at which a decision will be taken so that the full information is made available to all.

The second point is that the process is largely informal and that decisions are taken within an atmosphere of trust.

The third point is that the engineer in the head office is used as a resource which the FTL can call upon.

The FTL has other responsibilities in terms of integrating the work of all separate divisions who might be operating in the field, such as technical, financial, and community workers. He is not, and should not hold himself out to be, a specialist in any of these fields, so he cannot direct people regarding the technical aspects of their work. However, he can ensure that staff discipline is maintained; that all staff know what is going on and give the same message to members of the public; and that they collaborate effectively with each other.

Thus while the old silo concept of divisions based on areas of professional expertise survives, different professions collaborate in a system that integrates them into a single team. This can create conflicts in terms of supervisory responsibilities unless the roles are very clearly defined, but with practice can work extremely effectively. In Lusaka it was found that members of the team were the best form of control in that they used social pressure to discipline those who turned up late for work, or who gave the team a bad name by being rude to members of the public. This team-building approach was used consciously as a management tool, and it worked.

Another alternative which worked well in the Kecamatan Development Project in Indonesia was the use of special ad hoc staff for the project. Some 13,500 village-level facilitators were hired and trained—one facilitator for 650 people. They reported to consultants (a total of 1,200, reporting to 16 private companies) who provided technical assistance and guidance to the project at national, provincial, district, and Kecamatan levels. This was possible because the project became embedded as a way of doing things, so it was possible to develop a strong training program and have clear operational guidelines which provided a framework which was easy to replicate at the local level.[18]

COMMUNITY CONTRACT MANAGEMENT

The experience of Indonesia is one of the nearest to the concept being advocated here. The Kampung Improvement Program (KIP) was started in 1969, and has evolved over the years to become the largest and one of the more successful community-based development initiatives in the world. Some of the lessons learned in the KIP were used as the basis for the Kecamatan Development Project referred to above. This covered more than 30 percent of the country's rural subdistricts, and was based on the concept that villagers themselves would undertake the project planning and development and would receive and administer funds for development of projects such as road building and water schemes. Community members could be engaged both as paid laborers and in a voluntary capacity (to make the funds go further). As was mentioned in Chapter 10, there was potential room for the misuse of the funds, but methods for minimizing these risks were developed.

> The project's objective is fundamentally radical: it aims to institute transparency and democracy from the bottom up in a country where serious abuse of office and top-down planning have been endemic. It calls on villagers to demand accountability from both the government and their neighbors, and to take responsibility for the investments they deem important. This is a highly ambitious, indeed downright subversive, objective and one that is very risky in terms of program sustainability. The program will only achieve and sustain its objectives if people are able to insist on exercising and retaining this voice in decisions that affect them.[19]

There are three characteristics here which predispose the project to success:

- The communities are relatively stable, and have a history of self reliance and mutual aid.
- The amounts concerned are quite small.
- Improvements are incremental.

Box 11.2 The KIP program—Indonesian or universal lessons?

When I was teaching at the Boucentrum International Education (now Institute for Housing and Urban Development Studies), Rotterdam, we were invited to run short courses to train the KIP staff from the hundreds of local governments involved. At the launch of the first course, the Director of the Bandung Institute of Technology – speaking in Bahasa Indonesia – described the importance of the course, and the value that it would bring to the thousands of participants who would participate over the years. Then he introduced me. "He has a lot of experience in Africa, but what can we learn from Africa?" he said. Whether he realised that my counterpart was translating I do not know. He may, of course, have been right – in one sense conditions in different countries do require different solutions. But I also know he was wrong: different continents have much to teach each other. The essence of Indonesia's success in this program can be applied in all cultures and in all countries. (RM)

DELEGATED SPENDING POWERS

A very different example comes from Mozambique. Schools were traditionally managed in a top-down manner by the Ministry of Education: this management included the budgeting, management of teachers, and so on, with a supervisory role being played by provincial and district directorates of education. It was decided that this model was not working, and instead individual schools should operate their own budgets. The objective was to increase the efficiency and effectiveness of education, to redistribute political power among central, provincial, and district institutions, and individual schools; to improve the quality of education; and to change the way schools structured the learning environment and governed themselves. The local communities were brought in to enhance transparency, responsiveness, legitimacy, and accountability.

Clearly, this was a major shift in both policy and practice and represented an important shift in terms of powers. As an example of the new balance of power: instead of the ministry deciding which materials to buy, this was done by the school council. Once this decision was taken, the purchase process was undertaken locally. The school director and a community representative went first to the district directorate to receive a check to cover the costs and then to the bank to withdraw the money. They filled in two copies of the order note, leaving one copy and invoices from purchases with the district directorate, and keeping the other copy for the school.[20]

Despite some limitations, the project's impact on governance has been substantial and constructive in terms of building capacity, promoting participatory mechanisms, and changing attitudes.[21] Assessing the lasting impacts on learning conditions and pedagogical achievement is difficult at this point, but there have been higher rates of students passing key lower and upper primary school examinations, a substantial reduction in dropout and repetition rates and increase in gross enrollment rates, and higher rates of girls' enrollment and retention. Also, interestingly enough, the availability of funds in remote primary schools is stimulating development of the informal sector, which can contribute to reducing absolute poverty there.[22]

PROBLEMS

Community Funding

Community control needs to be managed carefully: that much is clear. It is not easy to change a system overnight, either from the perspective of a bureaucracy or from the community's side. Good ideas at the wrong time can run into trouble. In a small way, this system was established in South Africa for housing projects (see Box 11.3). The principle was that communities would have access to housing grants and subsidies from the state as a matter of right, provided that certain conditions were met. Unfortunately, the ground was not

Box 11.3 The Community and Urban Services Support Project.

USAID devised a creative system for helping poor communities to access grants and mobilise themselves. Once a community had been identified, and the project staff had—usually with the help of a local NGO—decided that it was sufficiently stable and well organized, the project helped the community to form a Trust with support from a legal firm whose fees were paid by the project. Once a Trust was formed it could be given funds (normally about $30,000) to engage technical experts to prepare a housing project for their area. The project itself would be fully grant-funded by the state, but the funds were not available until elaborate project preparation procedures had been complied with. In practice the Trusts worked well: what was less satisfactory was the rigidity of the national subsidy system, and the planning procedures required.

Source: Lance Bailey and Associates: *Housing and Community, an Overview of the Community and Urban Services Support Project* (Johannesburg: Lance Bailey and Associates, 1996), p9.

ready for this type of very creative relationship at any scale. On the one side the communities had unrealistic expectations about their powers and rights, and the militant and confrontational politics of the past spilled over into project design and development. The situation was exacerbated by a lack of community involvement skills within central and local government, and, more importantly, a very weak and rapidly evolving local government system.

Partnership Difficulties

Some of the experiences of Latin America have brought out potential difficulties:

When coordination with other parties is required, there emerges a host of problems in adapting internal procedures to those of others. Most government agencies are used to working in isolated, monolithic fashion. Communication among different parties is made difficult by the lack of fluid channels of communication . . .

The private sector tends to expect shorter time frames than does the public sector for execution of any project phase, which can lead to particularly severe problems. Tensions can also arise out of the dual requirements for control by the state on the one hand and flexibility on the other. A common complaint from nongovernmental organizations (NGOs) is that they are forced to account for the minutiae of their expenditures when working under contract for a state body in a way that public bodies are not required to do, and that they thereby waste time and energy. Compounding these frustrations are issues of slow disbursement—and these are problems that arise when the parties involved are trying their best to work together.[23]

Box 11.4   Mixed experiences from Rwanda.

I did a study of the management of clinics in Rwanda. The concept was that although the clinic staff were paid by the Ministry of Health, the money for drugs would come from payments made by the patients. Thus each clinic had a float which was used to purchase drugs from the central stores which was replenished by the patients' payments when they bought them. Each clinic was placed under the management of a local committee, thus placing the clinic under the scrutiny of the local community. However, the system broke down very quickly, because the Ministry of Health did not supply the drugs when they were ordered, thus leaving the clinics without the tools, so to speak, of their trade. As a result, people stopped even trying to use them, and on our visits there many of the clinics were empty of patients. Instead, the patients were willing to travel long distances to visit mission-managed clinics which they knew would have the drugs and supplies to provide the treatment. Thus it was that the partnership concept, which sounds so attractive, collapsed due to a failure by the public sector to keep its part of the bargain. By contrast, mission-funded clinics were working well and were highly popular. (RM)

Partnerships only work where both parties keep their side of the bargain. The collapse of the clinic system in Rwanda (see Box 11.4) vividly illustrates the problem where this does not occur.

By contrast, here is a positive example of a system which works. Concerning his days when he was director of the World Bank office in the Ivory Coast, Robert Calderisi wrote:

One day in the early 1990s, I visited a number of clinics outside the capital of the Ivory Coast. Most of them were filthy, dilapidated, and bare—lacking even syringes, bandages, and medicines. The personnel were idle, listless, and for obvious reasons, demoralized. In one clinic a mother had been torn during a difficult birth, and the midwife had been forced to mend her with electrical copper wire. By the end of the morning I too was disheartened. But, before returning to town, I passed by another clinic where the nurse met me at the door. She was well-dressed, clean, and plainly happy to have a visitor. The place was well-equipped and in good order. She had medicines, alcohol, and bandages. Her furniture was still usable. Her notebooks for patient follow-up were impeccable. And she was in the process of replacing wooden doors and windows that had been eaten away by termites.

What made the difference? Short of alternatives and unwilling to wait for government, the nurse had convinced the women of the village to contribute to the running of the clinic. She had no water, so she had contacted an Italian charity to repair the well. She had the conviction, energy, imagination, and charm to compensate for the lack of normal structures. Thinking that perhaps she had just arrived and that her enthusiasm was understandable, I asked how long she had been there. Eleven years, she told me.[24]

There are two lessons in this story. The first is the need for partnership and collaboration in service provision. The second is the need, as has also been repeatedly pointed out, for continuity and patience. This is a case of genuine development.

Loss of Power

Any change in the way that development is handled can meet with resistance by vested interests, the most of important of which is the bureaucracy. This resistance is likely to be even more energetic if power is transferred from the bureaucracy to the community. Sometimes the resistance will be more passive than active, such as delays in taking decisions, making payments, and appointing staff. A good example comes from the Northern Uganda Social Action Fund (NUSAF) project. The task was to:

> empower communities in Northern Uganda by enhancing capacity to systematically identify, prioritize, and plan for their needs and implement sustainable development to improve socioeconomic services and opportunities. In so doing, NUSAF will contribute to improved livelihoods by placing money and its management in the hands of the communities.[25]

The project was located within the office of the prime minister, and had substantial funding. Comprehensive implementation guidelines were prepared. But because none of the project design team members were seconded or recruited to the management unit, there was a disjuncture between design and implementation in respect of staffing. It took eighteen months to appoint district technical officers, and the two main directors responsible for the three main development components were not appointed until about two years after the project started. None of the district technical development officers had received vehicles almost two and one half years later, and only a few had been given computers or furniture. Similar difficulties affected disbursements: communities were required to open accounts in their local banks into which funds were directly disbursed in three tranches. Theoretically, transfers took just twenty-four hours, but delays between subproject approval by the management unit and actual disbursements meant that they took several weeks.[26] As Manor puts it:

> One key task is to persuade governments that new initiatives are non-threatening. Some of the most impressive achievements noted in our case studies occurred when governments concluded that changes

initially appearing to imperil them could actually enhance their effectiveness and responsiveness and thus their legitimacy and popularity. Government leaders—especially senior politicians who are almost always the key figures within them—are especially likely to feel threatened when they are asked to part with certain powers. In seeking to persuade them to do so, reliance on experiments within a small number of arenas may be crucial.[27]

Similar considerations applied in the Kecamatan Development Project. At the top level, government recognized that corruption and a sticky bureaucracy were impediments to successful development, and that it was essential to bypass these layers if the project were to succeed.

> The target population's perception of government and, possibly more important, the government's own perception of its credibility with this population, were so negative that the government was willing to experiment with a radical, new approach. KDP provided for funds to flow directly from a central project account to a joint village account at a local sub-district bank, processed by the branch office of the national Treasury.[28]

However, it seems that in this case, the obvious achievements of the project have been sufficient to compensate for the loss of power by government staff:

> Government officials have supported KDP in the hopes of benefiting from the associated political capital, even though they have no access to the funds themselves.[29]

MAKING CHANGE WORK

There is no magic wand in these situations, but it is possible to identify certain aspects which will make change more likely to succeed. Probably the most important component is that the staff involved in any such development must understand their role and must believe in it. These are the barefoot bureaucrats described in Chapter 8, whose role is one of supporters rather than controllers. This will only happen if there is a deliberate program to help people understand this new way of working.

Nabeel Hamdi describes this process well.

> The first visits of the expatriate team were, significant, about changing the hearts and minds of ministers and bureaucrats in favor of what became known as a 'support-based approach' to housing and urban development. What did it mean? Whose authority would it threaten?

What would it entail on the ground and how would it get started? . . . With the help and motivation of entrepreneurial individuals with status and power in government positions, we set a new trend which would give substance to the enabling approach . . . It began by cultivating first the right political climate for it all to happen on the ground, for liberating the latent potential of the everyday, and then by ensuring that what happened on the ground helped shape and redefine the politics and governance of city places.[30]

Innovations such as this work better if they can be built on existing structures and systems, and changes introduced on an incremental basis. In others words, change will, paradoxically, often take place more effectively and quickly if it does not require a total revolution in traditional roles and responsibilities. It is always tempting to throw out all the old ways of doing things, but by so doing we may create a vacuum in terms of people not knowing what they should do, and how they should do it.

Donors are often tempted to solve all the challenges and inefficiencies by a comprehensive review of all parts of the system.

> Donor agencies tend to favor ambitious reforms, but caution is advisable . . . some initiatives are ambitious in terms of scale—as when a program in northern Uganda that we analyzed sought to bring changes to seven development sectors and became overstretched.
>
> Other initiatives are ambitious in terms of degree or the magnitude of changes sought. They involve fundamental reforms entailing decisions that challenge formidable interests. Many governments in developing countries hesitate to attempt such initiatives, because they carry

Box 11.5 South Africa's penchant for comprehensive reform: A cautionary tale.

The apartheid system of local government was based on different local government units for different race groups. Clearly this had to be abolished, but it was not until six years after majority rule had been achieved that the new framework was agreed. In the transitional six years, the boundaries of black and white local authorities were changed to constitute integrated units, but this was not enough: all legislation relating to local government management was scrapped, and all boundaries were changed again. Thus, overnight, a totally new system was introduced. I was working closely with local government at the time, and could only admire the managers of some local government units, which had found themselves covering an area of which the boundaries were 100 kms away in all directions, with no offices, no telephones, and no staff. Even the Municipal Manager of one such local government unit was only a temporary secondment from the province. Even today many local governments have been unable to fulfil their mandate due to a lack of skills and funding. (RM) serious political risks. They often prefer to undertake more modest, incremental reforms that do not produce macrosystemic change, because their capacity and legitimacy are open to serious doubt, and because they find it more difficult than well-entrenched governments to withstand reactions from potent interests.[31]

At the project level too, incremental change seems to work better than the big-bang approach.

> Urban upgrading is always complex, and it is good practice to achieve scale through a series of incremental steps, which involves learning by doing, rather than trying to start a program already at scale.
>
> Low income communities usually have already been building their settlement for years, and some form of informal community upgrading project already exists. By giving itself a little time to learn how best to relate to its target communities and coordinate the actions of its stakeholders, a program increases its own chances of success.[32]

In other words, a gradual strategic transformation of the development mechanism works best: the long-term gradualistic partnership working towards a comprehensive transformation of the power relationships between the actors.

SOCIAL CAPITAL AND DEVELOPMENT

Social capital is a term that has gained much currency in the last ten years, and it has been a useful concept in terms of the ability and willingness of a community to take the lead in matters of common interest and to have the capacity to take responsibility.

Much of this book makes assumptions about the store of social capital which exists in informal settlements. There are many reasons why such settlements have relatively more social capital than other communities, of which the most important is hardship. People weave a network of interdependency as a result of hardship and economic stress. Social capital is also the product of external threats: it is remarkable how the prospect of some threat to the present order can bring people together. In the middle classes it can be the threat of a nuclear power station in their neighborhood, or a new road which will destroy a valuable and much loved forest. For the poor, it can take the form of the destruction of their community. Whatever the cause, threats generate social cohesion and a feeling of mutual dependency and trust.

But caution is needed about two aspects. The first is that in some communities social capital may be insignificant or nonexistent. There are many reasons for this, including the sheer stress of individual poverty and the concomitant need to devote all free time to earning a living.

In great cities men are brought together by the desire of gain. They are not in a state of co-operation, but of isolation, as to making of fortunes; and for all the rest they are careless of neighbors. Christianity teaches us to love our neighbor as ourself; modern society acknowledges no neighbor.

(Benjamin Disraeli, 1845).[33]

Other communities are populated by a fast-changing class of tenants who have no stake in the settlement, other than as temporary occupiers. They are obviously concerned about the standard of infrastructure such as water and roads but, having no financial stake in the settlement or long-term connection with it, tend to look at community activism as irrelevant. The landlords, who own the buildings, may be little interested in investing in improvements, and even opposed to formalization of the settlement because it will expose the extent of their property holdings and the exploitative levels of rents to public scrutiny.[34]

This is a major problem in some settlements and it requires a different approach, which can take time. However, even though they are tenants, many residents of such communities will eventually play a bigger role. NGOs can play a very effective role here, in terms of advising the tenants in terms of their powers and the treatment they receive, so that they form a cohesive group which can work with the landlords and other interest groups within the community.

Thus there is no standard solution, but rather a set of principles which can be applied. Each community will require different approaches and will develop different answers.

The second aspect concerning which caution must be expressed is that it is easy to view communities through, as Robert Putnam puts it, "gauzy self deception"[35] about the power and value of community activism. We would be the last to see the proposals being discussed here as being in any way a romantic view of development. Community involvement is a school of hard knocks, and only the experienced can deal with these knocks without relapsing into confrontation and distrust. Our sense is that politics, especially international politics, is increasingly polarized, and this is, unfortunately, rubbing off onto local politics and development. The title of a book (that has proved disturbingly prophetic in describing the potential for conflict in the world) can also be used as a descriptor of the potential for conflict between the "order" of the conventional city and the "disorder" of informal settlements: The Clash of Civilizations.[36]

Community-driven projects are not the easy way out, and the communities of informal settlements have within them the same potential for division as any social grouping, large or small. But this is not the point. Our point is simply that the communities of the poor deserve the same treatment as the rich: they have a right to decide for themselves what they want and how much they can pay for it. The reluctance of governments and development agencies to respond to this need has been born out of a mixture of caution and ignorance. Caution, because for some the risks are too great, and ignorance because they simply cannot conceive of a system that allows such relationships.

We have tried to suggest that the rewards of structuring development around the needs and capacity of the residents are not such a complex task, and although it presents unfamiliar challenges it is the route by which lasting and effective development can be achieved.

Notes

NOTES TO THE FOREWORD

1. Tony Gibson is the originator and developer of Planning for Real, the most widely used participatory planning method—see www.nif.co.uk.

NOTES TO THE INTRODUCTION

1. UN Habitat, *Slums of the World: The Face of Urban Poverty in the New Millennium?* (Nairobi: United Nations Human Settlements Programme, 2003), 10, 33.
2. Paul Collier, *The Bottom Billion: Why the Poorest Countries Are Failing and What Can Be Done about It,* (Oxford University Press, 2007), 7.
3. Noni Jabavu, *Drawn in Colour,* London 1960, p. 116. Quoted in P. T. Bauer, *Dissent on Development* (Cambridge, MA: Harvard University Press, 1972), 106.
4. P. T. Bauer, 103.
5. Alain de Botton, *The Architecture of Happiness* (London: Penguin, 2006), 2.
6. Amartya Sen, *Development as Freedom* (Oxford, Oxford University Press, 1999).
7. Sen, 285.
8. Sen,17.
9. For example, national and local governments, but development assistance agencies—as Chapter 4 shows—are not exempt from this procedural temptation.
10. Arjun Appadurai, "The Capacity to Aspire: Culture and the Terms of Recognition," in *Culture and Public Action,* Vijayendra Rao and Michael Walton (eds.), Stanford, CA: Stanford University Press, 2004.
11. John F. C. Turner and Robert Fichter (eds.), *Freedom to Build* (New York: Collier Macmillan, 1972); John F. C. Turner and Robert Fichter, eds., *Housing by People* (London: Marion Boyars, 1976).

NOTES TO CHAPTER 1

1. Specifically it required all new residential construction to include running water and an internal drainage system.
2. Jacob A Riis (1890): *How the Other Half Lives.* Reprinted by Dover 1971, New York, 5–6.

3. Under the British Town and Country Planning Act of 1947. This act, in a sense, gave the state the right to acquire compulsorily and use all land as it deemed fit, subject to the payment of fair compensation.

4. The word was first used by Richard Jolly to describe traders who were operating without licenses and who thereby avoided the expense, delays, and complexity of complying with the requirements of operating a formal business. Since then it has been widely used in connection with urban development—as here.

5. Malcolm Gladwell, *Blink: The Power of Thinking without Thinking* (London: Abacus, 2006).

6. Gladwell, 189.

7. Under the Millennium Development Goals, a slum household is defined as one that lacks any one of the following five elements: access to improved water, access to improved sanitation, durability of housing, sufficient living area, and security of tenure. In practice, it is the first four criteria that are used.

8. Malcolm Gladwell, *Blink: The Power of Thinking without Thinking* (London: Penguin, 2006).

9. Malcolm Gladwell, *The Tipping Point: How Little Things Can Make a Big Difference*, Penguin, 2002.

10. Neuwirth, Robert (2005), *Shadow Cities, A Billion Squatters, a New Urban World*, Routledge, New York, 119.

11. Solomon J. Greene, *Staged Cities: Mega-Events, Slum Clearance, and Global Capital,* Yale Human Rights and Development L.J., Vol. 6, 2003 (from http://islandia.law.yale.edu/yhrdlj/PDF/Vol%206/greene.pdf).

12. In Lagos, the inner-city housing was labeled a "disgrace to the capital city" in the Economic Programme, and "a humiliation to any person with a sense of national pride" by the minister of Lagos affairs. See Marris, Peter (1961): *Family and Change in an African City: A Study of Rehousing in Lagos* (London: Routledge and Kegan Paul), 119.

13. In the very informal layout of squatter settlements in Zambia it was possible to demonstrate that sewerage (typically the service which is the most difficult to retrofit) was cheaper in existing settlements than new ones, because the densities were significantly higher (Paul Andrew, Malcolm Christie, and Richard Martin: "Squatters and the Evolution of a Lifestyle," *Architectural Design*, Vol. XLIII, 1/1973), 16–25.

14. In the early 2000s, for example, the Philippines' government was resettling squatters into units averaging about 20 m², whereas a survey conducted (by Mathema in 2002) in the informal settlements showed that the average built-up area being used by the beneficiaries in their original place of residence was in the range of 40 m² (not counting the external open space). This continues to be a common practice in many countries.

15. Marris, Peter (1961): *Family and Change in an African City: A Study of Rehousing in Lagos*. London: Routledge and Kegan Paul, 91.

16. Nehru, Jawaharlal (1958), in a foreword to Bharat Sevak Samaj: *The Slums of Old Delhi*. Delhi, Arma Ram and Sons (quoted in Marris, Peter, 116).

17. Marris, Peter (1961): 124.

18. Michael Young and Peter Willmott, *Family and Kinship in East London* (London: Penguin, 1957).

NOTES TO CHAPTER 2

1. The information presented here is based on field research in these cities and informal settlements conducted by Mathema in: the Philippines (2002), Afghanistan (2003), Swaziland (2003–04), Ethiopia (2004), Eritrea (2005),

Kenya (2005), China (2005–06), Mongolia (2006), Tanzania (2006), Ghana (2006–07), Nigeria (2008), and UAE (2007–08).
2. In the case studies and subsequently wherever the symbol $ is found it refers to U.S. dollars.
3. Robert Buckley and Ashna Mathema, "Real Estate Regulations in Accra, Ghana: Some Macroeconomic Outcomes," *Urban Studies,* August 2008; and Robert Buckley and Ashna Mathema, "Is Accra a Superstar City?" *Policy Research Working Paper* 4453, World Bank, December 2007.
4. Ashna Mathema, *Housing in Addis: Background Paper, 2005* (unpublished background research for the World Bank, funded by the Danish Trust Fund).
5. Ashna Mathema, *Qualitative Study: Household Interviews, Dar es Salaam, Tanzania,* 2007 (unpublished background research for African Union for Housing Finance, funded by Cities Alliance/the World Bank).
6. Ashna Mathema, *Slums and Sprawl: The Unintended Consequences of Well-Intended Regulation,* 2008 (unpublished background research for the World Bank).
7. Metropolitan Household Survey, Lagos Central Office of Statistics (COS), 2005.
8. LMDGP Project Appraisal Report, 2006, World Bank.
9. Ashna Mathema, *Nairobi's Informal Settlements: A View from the Inside,* 2005 (unpublished background research, funded by the Norwegian Trust Fund/the World Bank).
10. The lack of latrines and the dangers of going out at night have led to the practice of using a plastic bag, which is then tied up and deposited on a dump, or simply thrown outside—hence the term "flying."
11. Marie Huchzermeyer, *Today's Tenement Cities: Large Scale Private Landlordism in Multi-storey Districts of Nairobi, Draft Research Report* (Johannesburg: University of the Witswatersrand, 2006), 11.
12. Marie Huchzermeyer, 4.
13. PADCO, *Metro Manila Urban Services for the Poor Project*, Asian Development Bank, 2003. The field work and housing/socioeconomic assessment of Manila's slums under this project was led by Mathema in 2002.
14. A common arrangement is for anywhere between five and fifteen persons sharing a single room. The rooms have bunk beds (as many as five to six bunks in one 15–20 m² room), and often the bed is used by two or three people in different shifts. There is a common cooking area and toilet assigned to each room.
15. Taken from *Gulf News*, properties section, April 11, 2009.
16. Roger Hardy, "Migrants Demand Labour Rights in Gulf," BBC News, 27 February 2008, http://news.bbc.co.uk/2/hi/middle_east/7266610.stm.
17. There are also some camps exclusively for women labor, but much fewer in number. The vast majority are for men.
18. "*Panorama: Slumdogs and Millionaires*," BBC One, broadcast on 6 April 2009 (http://www.bbc.co.uk/iplayer/episode/b00jqgww/Panorama_Slumdogs_and_Millionaires).

NOTES TO CHAPTER 3

1. These interviews were conducted by Ashna Mathema for the World Bank as part of separate assignments to analyze housing and socioeconomic conditions in informal settlements in: Ethiopia (2004), Swaziland (2003–04), Kenya (2005), Tanzania (2006), Ghana (2006–07). The interviews documented

here are a small part of a larger sample, which can be found in the following unpublished documents by Ashna Mathema:

Housing in Addis: Background Paper (unpublished research for the World Bank, funded by the Danish Trust Fund, 2005).
Mbabane Upgrading and Finance Project: Household Interviews unpublished research for Mbabane City Council, funded by Cities Alliance/the World Bank, 2005).
Nairobi's Informal Settlements: A View from the Inside, background paper (unpublished research funded by the Norwegian Trust Fund/ the World Bank, 2005).
Qualitative Study: Household Interviews, Accra, Ghana (unpublished research for African Union for Housing Finance, funded by Cities Alliance/the World Bank, 2006).
Qualitative Study: Household Interviews, Dar es Salaam, Tanzania (unpublished research for African Union for Housing Finance, funded by Cities Alliance/the World Bank, 2007).

NOTES TO CHAPTER 4

1. World Bank, *World Development Report, 2009, Reshaping Economic Geography* (World Bank, Washington, DC, 2009), 68.
2. Jane Jacobs, *The Economy of Cities* (New York: Vintage, 1970), quoted in *World Development Report, 2009*, 49.
3. HOPE VI projects in the United States are an example of this.
4. This refers to the medium-rise, 4–5 story walk-ups reminiscent of the failed public housing of the 70s, intended to densify while making it cost effective by eliminating the need for elevators.
5. Robert Buckley and Ashna Mathema, *Shelter Assistance for the 'Bottom Billion': Reconsidering Bank Policies*, draft manuscript, September 2007.
6. Ashna Mathema, *Nairobi's Informal Settlements: A View from the Inside*, background paper, 2005 (unpublished background research funded by the Norwegian Trust Fund/the World Bank).
7. Field research by Ashna Mathema in 2003.
8. Ashna Mathema and Nayana Mawilmada, *Decentralization and Housing Delivery: Lessons from the Case of San Fernando, La Union, Philippines*, unpublished thesis for Master of City Planning program at MIT, 2000.
9. Field research by Ashna Mathema in 2004.
10. Field research by Ashna Mathema in 2004–05.
11. Field research conducted by Ashna Mathema, 2002–03, as part of an exercise for the World Bank and the government of Swaziland to develop settlement profiles of fifty-two peri-urban areas which were to include housing and physical infrastructure conditions.

NOTES TO CHAPTER 5

1. Martin was then working for USAID's Regional Housing and Urban Development Office, which was offering a guaranty to Zimbabwe under the housing guaranty program to allow it to borrow funds for housing.
2. *World Bank: World Development Report, 2009* (Washington, DC: World Bank, 2009), 205.
3. S. Yahya et al., *Double Standards, Single Purpose* (London: ITDG, 2001), 105.

4. A. L. Mabogunje; J. E. Hardoy, and R. P. Misra, *Shelter Provision in Developing Countries SCOPE Report 11* (Chichester, UK: John Wiley & Sons, 1978), 45.

5. A. L. Mabogunje et al., 45.

6. There are many assumptions in the characterization of the process here. For example, we assume that the subject of this imaginary scenario has lived in town (typically renting) for sufficient time to save up some money for house construction, and has the support of family and community members in the land invasion and house construction process. Much the same procedures apply if the person's access to land is through informal subdivision, which is increasingly normal at the present time.

7. Thomas Lundgren, Ann Schlyter, and Thomas Schlyter: Kapwepwe Compound, 'A Study of an Unauthorised Settlement,' Lund, 1969. Mogens Christensen: unpublished survey, partly reproduced in *Government of the Republic of Zambia: Lusaka Sites and Services Project, request to the International Bank of Reconstruction and Development for a Loan* (Lusaka: National Housing Authority, 1974). 1975 survey by Richard Martin (unpublished).

8. Room sizes measured from Lundgren, Schlyter, and Schlyter: The average size from this study was 5.96 m² compared with the official minimum of 8.4 m².

9. To remove a bicycle from on top of a galvanized iron roof without the occupants of the room underneath hearing is virtually impossible, so this is a very secure system.

10. National Housing Authority: *Criteria for the design of low cost and very low cost houses,* Lusaka 1972.

11. This sounds very small, but is in fact only 1 percent less than the 1/20th of the floor area required under the Public Health Act.

12. In some cities, the people whose interests are being enforced are minorities in the cities concerned, and the occupants of informal housing are in the majority. It is ironic that this privileged minority should have the power and capacity to enforce their rules in this way.

13. Hernando de Soto, *The Other Path: The Invisible Revolution in the Third World* (New York: Harper and Row, 1989), 132–51.

14. RM: personal experience.

15. Ekurhuleni Metropolitan Council, *Upgrading for Growth: Generic Estimate of Project Implementation Activities* (unpublished, 2007).

16. World Bank, *Housing: Enabling Markets to Work* (Washington, DC: World Bank Policy Paper, 1993), 85.

17. In Swaziland, for example, houses may be constructed using the traditional pole and mud construction under the Grade 2 building regulations.

18. Robert Buckley and Jerry Kalarickal, *Thirty Years of World Bank Shelter Lending: What Have We Learned?* (Washington, DC: World Bank, 2006), 36.

19. Warren Smit, "Understanding the Complexities of Informal Settlement: Insights from Cape Town," in *Informal Settlements—A Perpetual Challenge?* Ed. Marie Huchzermeyer and Aly Karam (Cape Town: UCT Press), 122.

20. Mabogunje et al., 80.

21. There is potential for development in this area to avoid complex legislation. One of the interesting features of the township legislation in South Africa is that each developer can, within limits, determine the standards of the development (such as road widths, plot sizes, etc.). A township can consist of anything from a few (say twenty) plots right up to a very large development of many thousand plots.

22. Government of the Republic of Zambia: Housing (Statutory and Improvement Areas) Act (Lusaka: Government Printer, 1974).

23. Richard Martin, "Institutional Involvement in Squatter Settlements," *Architectural Design*, Vol. XLVI (April 1976), 232–37.

24. Apollo Njonjo, *Study on Community Managed Water Supplies, Draft* (Nairobi: Report on Case Studies for the World Bank Regional Water and Sanitation Group—East Africa, UNDP/WB Water and Sanitation Programme, 1994), quoted in Deepa Narayan, *Designing Community Based Development* (Rome: SD Dimensions, UN Food and Agricultural Organisation, Sustainable Development Department, 1997).

25. Robert Neuwirth, *Shadow Cities: A Billion Squatters, a New Urban World* (New York: Routledge, 2005).

26. Ashna Mathema, *"Planning and Practice: The Missing Link"* (paper prepared for the International Conference on Homelessness: A Global Perspective, Delhi, January 2007) (unpublished), 7.

27. Guyo Haro, Godana Doyo, and John McPeak, "Linkages between Community, Environmental and Conflict Management: Experiences from Northern Kenya," *World Development*, Vol. 33, No 2, 2005: 285–99.

28. That is to say, the practice within the community concerned.

29. PADCO: *Study on Upgrading in the Peri-Urban Areas of Swaziland*, prepared for the World Bank and the Ministry of Urban Development and Housing (Washington, DC: PADCO, 2003).

NOTES TO CHAPTER 6

1. Thomas Jefferson, letter to William Charles Jarvis, September 28, 1820; quoted in Nabeel Hamdi, *Housing without Houses* (Rugby, UK: Intermediate Technology Publications, 1995), 75.

2. Nick Graham, "Informal Settlements Upgrading in Cape Town: Challenges, Constraints and Contradictions within Local Government," in *Informal Settlements—A Perpetual Challenge?* Ed. Marie Huchzermeyer and Aly Karam (Cape Town: UCT Press), 243.

3. Amartya Sen, *Development as Freedom* (Oxford: Oxford University Press, 1999), 75.

4. Sen, 75–76.

5. Audit Commission: *Listen Up! Effective Community Consultation* (London: Audit Commission, 1999), 32.

6. Norman Schwartz and Anne Derruyterre, *Community Consultation, Sustainable Development and the Inter-American Development Bank* (Washington, DC: IDB, 1996), 5.

7. Schwartz and Deruyterre, 2.

8. Schwartz and Deruyterre, 3.

9. Parmesh Shah (2000), "Participation in Poverty Reduction," *North-South Institute, e-views*, Issue 7, 3.

10. The work of Leon Festinger is particularly useful in this context, and, although it was published many years ago, it has lost none of its relevance. Two works are especially relevant: Leon Festinger, *A Theory of Cognitive Dissonance* (Stanford, CA: Stanford University Press, 1957); and Douglas H. Lawrence and Leon Festinger, *Deterrents and Reinforcement* (Stanford, CA: Stanford University Press, 1961).

11. Festinger, *A Theory of Cognitive Dissonance*, 52.

12. Festinger, 3.

13. Festinger, 78.

14. Festinger, 91.

15. Religious groups are the most obvious example of this; indeed, in Africa, religion is a very strong influence in informal settlements. However, the groups referred to here embrace every type of group, from football team to a mothers' group; from a trade union to a political party.

16. The concepts discussed in this section are incorporated in a live simulation game, Power, Participation and Protest, which can be downloaded from the book Web site: http://www.developmentpovertyandpolitics.com.

17. BBC Hardtalk, 19 December 2006.

18. Constitutional Court Case 2002 (1) SA 46 (CC). Irene Grootboom's fate described in *Mail and Guardian Online,* August 8, 2008.

19. An account of this is contained in a DVD published by Shack Slum Dwellers Association: Indian Alliance (Mahila Milan/National Slum Dwellers' Federation N.S.D.F/Society for the Promotion of Area Resource Centre (SPARC), "Seeing is Believing" (2004).

20. Nabeel Hamdi, *Housing without Houses, Participation, Flexibility, Enablement* (Rugby, UK: Intermediate Technology Publications, 1995), 75.

21. Readers will be aware of the many ways in which the traditional business model is distorted by monopolies (which by definition remove the option of choice), unfair and dishonest practices, and so on. We could examine this and the nature of contracts (oral and written) at length, but we prefer to concentrate on the characteristics and outcomes of business relationships in the broad sense of the terms.

22. There is an obvious overlap here between "participation as public relations" and the communication mentioned here. Public relations communications are designed to sell; project support communication is designed to give the audience/readership the information to make an informed decision. These are very different.

23. Alana Potter and Kate Skinner, *Lessons Learned for Improving Communication at the Local Level,* paper presented at the 11th ITN Africa Conference, Harare, Zimbabwe, 1999, 5.

24. Ashna Mathema, *Nairobi's Informal Settlements: A View from the Inside,* background paper (unpublished background research funded by the Norwegian Trust Fund/the World Bank), 2005.

25. Philip Zimbardo, *The Lucifer Effect: How Good People Turn Evil* (London: Rider, 2007), 172.

26. Janice E. Perlman, *Favelas of Rio de Janeiro, Urban Informality,* 130–31.

27. Perlman, 130.

28. Amartya Sen, *Development as Freedom* (Oxford: Oxford University Press, 1999), 268.

29. Robert Neuwirth, *Shadow Cities: A Billion Squatters, a New Urban World* (New York: Routledge, 2005), 274–75.

30. is shorthand in Nigeria for a crime of deception with the intent of committing fraud. The number refers to a clause in the criminal code of Nigeria which applies to this procedure. It was originally applied to e-mail fraud in which outsiders were offered huge sums of money if they would allow their bank accounts to be used to laundry funds which a prominent person or institution had available, but could not access.

31. Daniel Jordan Smith, *A Culture of Corruption: Every Day Deception and Popular Discontent in Nigeria* (Princeton, NJ: Princeton University Press, 2006), 226.

32. Barney Jopson, "Clean Break," *Financial Times Weekend, Life and Arts,* November 29, 2008, 1. For a fuller account of the process used, see R. Ahmed, *Achieving 100% Sanitation, WaterAid and VERC Approach* (Dacca: WaterAid, 2006).

33. Village Education Resource Center, *Shifting Millions from Open Defeca-tion to Hygienic Latrines, Process Documentation of 100% Sanitation Approach* (Dacca: VERC, 2002).

NOTES TO CHAPTER 7

1. C. K. Prahalad, *Fortune at the Bottom of the Pyramid: Eradicating Poverty through Profits* (Upper Saddle River, NJ: Wharton School Publishing, 2006), 17.
2. This was particularly strongly stressed in the one-party state systems of Tan-zania and Zambia, for example.
3. Indonesia offers a good example of this: the Kampung Improvement Programme relied strongly on existing Islamic self-help and mutual support groups.
4. Swapping houses may sound rather an odd concept, but it worked very well in Zambia, where people who were living in a congested area and who wanted larger plots, and were willing to build a new house, swapped with someone whose house was due to be demolished to make way for a new road. In that way, one family surrendered their unit which was to be demolished to move into an existing house, the owners of which had been given a new plot. Similar concepts were well received in Swaziland (Richard Martin, Ashna Mathema, et al., *Mbabane Upgrading and Finance Project* (Washington, DC: Cities Alliance, 2007), 130.
5. If we can be allowed a substantial deviation from the main theme, it is to warn that in community-driven projects the use of financial compensation creates the potential for substantial conflict between the residents and the compensating authority. On the one hand, if payments are too low they will achieve nothing but disgruntlement; if they are too high they will create per-verse incentives to attract compensation. In Swaziland, for example, the cost of compensation paid under an upgrading project, largely for destruction of trees and hedges, but also of a few houses, was greater than the cost of the improvements to the roads and water systems. The fact that the valua-tors were paid a percentage of the sums paid out might have been a factor in inflating the sums paid. Be that as it may, there is nevertheless a likeli-hood that if the community had been given the responsibility for handling these funds the distribution might have been very different. By contrast, in Zambia, no one received compensation, and there was only one serious com-plaint: a bar owner was in the way of the road. Knowing that there was no compensation if his property was demolished, he refused to move. The com-munity agreed with him—it would have been unfair for him to have lost such a valuable asset—so the route of the road was changed to allow him to stay.
6. Roger Fisher, Elizabeth Kopelman, and Andrea Kupfer Schneider, *Beyond Machiavelli, Tools for Coping with Conflict* (Cambridge, MA: Penguin, 1993).
7. Dale Carnegie, *How to Win Friends and Influence People* (New York: Pocket Books, 1982), 101.
8. Dale Carnegie, 122; quoting from *Bits and Pieces* (Fairfield, NJ: Economic Press).

NOTES TO CHAPTER 8

1. This is where the term *barefoot* is so useful in expressing another path. It carries with it the bottom-up undertone that is appropriate to the situation, even though, if you think about the term objectively, it sounds rather pejorative.

2. Unfortunately, many of the examples of bad construction are done by relatives, or people claiming to be relatives of the client. They use their consanguinity to suggest that they will be cheaper and more reliable: far too often the opposite is true.
3. C. K. Pralahad, 221–34.
4. R. J. Martin, *Self Help in Action* (Lusaka: National Housing Authority, 1975).
5. There has been great success in this field in South Africa, in the context of the water company installing meters for the first time in the townships. The motivation from the water companies was to reduce water wastage, and to improve the level of cost recovery. As a way on encouraging people to sign up for a meter, they undertook the plumbing repairs previously discussed free of charge.
6. Alex Counts, *Small Loans, Big Dreams: How Nobel Prize Winner Muhammad Yunus and Microfinance Are Changing the World* (Hoboken, NJ: John Wiley and Sons, 2008).
7. Shaffer, J., et al. (1974), "Models of Group Therapy and Sensitivity Training," 194–95; quoted in Shanahan, Peter (1976), *Report on Community Development Training Programme and Proposals for Future Programmes; Housing Project Unit, Lusaka City Council* (mimeo), 7–9.

NOTES TO CHAPTER 9

1. For a detailed account of the history and operations of the Grameen Bank, see Alex Counts, ibid.
2. Richard Martin and Ashna Mathema: *Housing Finance in Morocco: Does It Work for the Poor?* (DAI, USAID, Cities Alliance, Washington, DC, 2008).
3. Richard Martin et al., *Development of Appropriate Housing Finance Products to Support Upgrading Activities* (Washington, DC: Cities Alliance, 2008).
4. This model only works where there is a rental market, but in our experience there is a steady stream of young people and recent immigrants to the city for whom rental housing is the preferred choice.
5. The first scenario is taken from Jeffrey Sachs, *The End of Poverty: How We Can Make It Happen in Our Lifetime* (London: Penguin, 2005), 247. The second takes considerable liberties with the model; any economic errors are the authors'.
6. Robert Neuwirth, *Shadow Cities, a Billion Squatters, a New Urban World* (New York: Routledge, 2005), 44–45.
7. Paul Polak, *Out of Poverty: What Works When Traditional Approaches Fail* (San Francisco: Berrett-Koehler, 2008), 145.
8. C. K. Pralahad, 221–34.
9. Richard Martin et al., *Development of Appropriate Housing Finance Products to Support Upgrading Activities* (Washington, DC: Cities Alliance, 2008), 79.

NOTES TO CHAPTER 10

1. David Pasteur, *The Management of Squatter Upgrading: A Case Study of Organisation, Procedures and Participation* (Farnborough, UK: Saxon House, 1979).

2. M. B. Bamberger, B. Sanyal, and N. Valverde, *Evaluation of Sites and Services Projects: The Experience from Lusaka, Zambia*. World Bank Staff Working Papers, Number 548 (Washington, DC: World Bank, 1982).
3. Robert Ledogar, *Community Participation, Collective Self-Help and Community Development in the Lusaka Housing Project* (Lusaka [mimeo], 1979).
4. This was the person heading the IDRC/World Bank–funded evaluation team.
5. Richard Martin, "Experiences with Monitoring and Evaluation in Lusaka," in R. Skinner, J. Taylor, and E. Wegelin, *Shelter Upgrading for the Poor* (Manila: Island Publishing House, 1987), 149–69.
6. Martin, 164
7. Martin, 157
8. Carnegie, 150.
9. Vance, Irene, "The Community as Evaluators: Experience with Community Participation in Self-Build Projects in Managua, Nicaragua; in Skinner et al., 169–96.
10. Skinner et al., 179.
11. Skinner et al., 181.
12. Skinner et al., 183.
13. Marie Huchzemeyer et al., "Policy, Data and Civil Society: Reflections on South African Challenges through an International Review," in *Informal Settlements, a Perpetual Challenge?* ed. Marie Huchzermeyer and Aly Karam (Cape Town: UCT Press, 2006), 29.
14. Sinha, Rama (1978), *Three Jeers for Bureaucracy* (Delhi: Orient Longman Ltd.), 9.
15. A self-explanatory word coined by the author.
16. Sinha, 31.
17. Benjamin A. Olken, *Monitoring Corruption: Evidence from a Field Experiment in Indonesia*. NBER Working Paper 11753, National Bureau of Economic Research (Cambridge, MA: National Bureau of Economic Research, 2005), 5–6.
18. Paul Collier, *The Bottom Billion: Why the Poorest Countries Are Failing and What Can Be Done about It* (Oxford: Oxford University Press, 2007), 150.

NOTES TO CHAPTER 11

1. The term *state* is used here and following as shorthand for all public-sector investment, whether by local, state, national or federal government, public utilities, and all other publicly funded agencies.
2. The millennium development goals, for example, Target 7.C: "Halve, by 2015, the proportion of people without sustainable access to safe drinking water and basic sanitation" (for updated information on progress, see http://www.mdgmonitor.org/goal7.cfm).
3. Note the confusion in terms commonly used—the state "serves" its citizens—the civil service is living proof of that concept; but yet the monarchical term survives in that we are all considered "subjects" of the state.
4. It must be conceded that some of the long-term concessions and build-operate-transfer contracts have a stronger partnership character than short-term arrangements, but the state remains the initiator, designer, and funder of the arrangement, and the private sector enters into the arrangement at the will of the state.

5. F. Appiah et al., *Fiscal Decentralization and Sub-National Government Finance in Relation to Infrastructure Service Provision in Ghana* (Washington, DC: World Bank, 2000).

6. Constitution of South Africa Act, 1996, s214.

7. The experience of using formulae in both Ghana and South Africa has been mixed: for example, in Ghana it has an anti-urban bias, so that larger cities receive far less than most would consider their fair share; similar, but less serious criticisms, are reported in South Africa. Therefore once a formula has been adopted, it should be reviewed on a regular basis. This is an excellent role for classic evaluation techniques.

8. Alex Counts, *Small Loans, Big Dreams* (Hoboken, NJ: Wiley, 2008), 15.

9. Ivo Imperato and Jedd Ruster, *Slum Upgrading and Participation: Lessons from Latin America* (Washington, DC: World Bank, 2003), 79.

10. Mike Davis, *Planet of Slums* (New York: Verso, 2006), 76.

11. Robert M. Buckley and Jerry Kalarickal, *Thirty Years of World Bank Shelter Lending: What Have We Learned?* (Washington, DC: World Bank, 2006). Reference cited: Arjun Appadurai, "The Capacity to Aspire: Culture and Terms of Recognition," in *Culture and Public Action*, ed. Vijayendra Rao (Washington, DC: World Bank, Stanford University Press. 2004).

12. A vivid account of this is contained in a DVD published by Shack Slum Dwellers Association. Track 7: Seeing is Believing—Indian Alliance, Mahila Milan/National Slum Dweller's Federation (NSDF)/Society for the Promotion of Area Resource Centre (SPARC), 2004.

13. Nabeel Hamdi, *Housing without Houses* (London: Intermediate Technology Publications, 1995), 103.

14. While often seen as hotbeds of crime, informal settlements are often relatively safe, so this example may not be a good one in some countries. On the other hand, there are communities where the power of gangs has become so great that they effectively control the settlement. This may lead to relative safety for those who do not challenge the gangs, but it also leads to warfare between the different gangs and drug cartels (Janice E. Perlman, *Favelas of Rio de Janeiro, Urban Informality*, 130–31).

15. A fuller example of this was provided in Chapter 8, in connection with a discussion of barefoot planners.

16. To demonstrate that it can work, we use a real example which was tried under extremely demanding circumstances and fully survived the test. This is described following.

17. The management system and operations are fully described in David Pasteur, *The Management of Squatter Upgrading* (Farnborough, UK: Saxon House, 1979).

18. Judith Edstrom, *Indonesia's Kecamatan Development Project: Is It Replicable?* World Bank: Social Development Papers, Paper No. 39 (Washington, DC: World Bank, 2002).

19. Judith Edstrom, 2.

20. Fidelx Kulipossa and James Manor, "The Direct Support to Schools Program in Mozambique," in *Aid That Works: Successful Development in Fragile States,* ed. James Manor (Washington, DC: World Bank, 2007), 162–64.

21. Kulipossa and Manor, 188.

22. Kulipossa and Manor, 165.

23. Ivo Imperato and Jeff Ruster, *Slum Upgrading and Participation: Lessons from Latin America* (Washington, DC: World Bank, 2003), 233.

24. Robert Calderisi, *The Trouble with Africa: Why Foreign Aid Isn't Working* (New York: Palgrave Macmillan, 2006), 224.

25. World Bank: *Project Appraisal Document on a Proposed Credit to the Government of Uganda for a Northern Uganda Social Action Fund, Report 23885* (Washington, DC: World Bank), 2.
26. Mark Robinson, "Community-Driven Development in Conflict and Post-conflict Conditions," in *Successful Development in Fragile States*, ed. James Manor (Washington, DC: World Bank, 2007), 275–78.
27. James Manor, ed, *Aid That Works: Successful Development in Fragile States* (Washington, DC: World Bank, 2007), 27.
28. Edstrom, 2.
29. Edstrom, 3.
30. Nabeel Hamdi, *Small Change: About the Art of Practice and the Limits of Planning in Cities* (London: Earthscan, 2004), 21.
31. Ivo Imperato and Jeff Ruster, 209.
32. Ivo Imperato and Jeff Ruster, 209.
33. Quoted in Robert Putnam, *Bowling Alone: The Collapse and Revival of American Community* (New York: Simon & Schuster, 2000), 380.
34. This problem is common, as discussed in Chapter 2, in the case of Nairobi and Addis Ababa.
35. This comment is part of the introduction to the above-referenced powerful book about social capital in the contemporary United States. His data seem to reflect a decline in community bonds since the 1950s, when civil society (and therefore social capital) was so strong. Robert Putnam, 25.
36. Samuel Huntingdon, *The Clash of Civilizations and the Remaking of World Order* (New York: Simon & Schuster, 1996).

Bibliography

Ahmed, R. *Achieving 100% Sanitation, WaterAid and VERC Approach.* Dacca: WaterAid, 2006.

Andrew, Paul, Christie, Malcolm, and Martin, Richard. Squatters and the Evolution of a Lifestyle. *Architectural Design*, Vol. XLIII, 1/1973, 16–25.

Appiah, F., et al. *Fiscal Decentralization and Sub-National Government Finance in Relation to Infrastructure Service Provision in Ghana.* Washington, DC: World Bank, 2000.

Arjun, Appadurai, "The Capacity to Aspire: Culture and Terms of Recognition." In *Culture and Public Action*, ed. Vijayendra Rao. Washington, DC: World Bank, Stanford University Press, 2004.

Audit Commission. *Listen up! Effective Community Consultation.* London: Audit Commission, 1999.

Bamberger, M. B., Sanyal, B., and Valverde, N. *Evaluation of Sites and Services Projects: The Experience from Lusaka, Zambia.* World Bank Staff Working Papers No. 548. Washington, DC: World Bank, 1982.

Bauer, P. T. *Dissent on Development, Studies and Debates in Development Economics.* Cambridge, MA: Harvard University Press, 1972.

Buckley, Robert, and Kalarickal, Jerry. *Thirty Years of World Bank Shelter Lending: What Have We Learned?* Washington, DC: World Bank, 2006.

Buckley, Robert, and Mathema, Ashna. *Shelter Assistance for the 'Bottom Billion': Reconsidering Bank Policies.* Draft manuscript, September 2007.

———. *Is Accra a Superstar City?* Policy Research Working Paper 4453. Washington, DC: World Bank, 2007.

———. "Real Estate Regulations in Accra, Ghana: Some Macroeconomic Outcomes." *Urban Studies*, August 2008.

Calderisi, Robert. *The Trouble with Africa: Why Foreign Aid Isn't Working.* New York: Palgrave Macmillan, 2006.

Carnegie, Dale. *How to Win Friends and Influence People.* New York: Pocket Books, 1982.

Carson, Lyn, White, Stuart, and Hendriks, Carolyn and Palmer. "Community Consultation in Environmental Policy Making." *The Drawing Board: An Australian Review of Public Affairs*, Vol. 3, No. 1, July 2002, p. 12.

Central Office of Statistics. Metropolitan Household Survey. Lagos: Central Office of Statistics, 2005.

Christensen, Mogens. Unpublished survey, partly reproduced in government of the Republic of Zambia. Lusaka Sites and Services Project, request to the International Bank for Reconstruction and Development for a Loan. Lusaka: National Housing Authority, 1974.

City of Johannesburg Metropolitan Municipality, Public Health By-laws. Johannesburg: *Provincial Gazette Extraordinary*, No. 179, 21 May 2004.

276   *Bibliography*

Collier, Paul. *The Bottom Billion: Why the Poorest Countries Are Failing and What Can Be Done about It.* Oxford: Oxford University Press, 2007.
Council for Scientific and Industrial Research. *Ghana: Agricultural Research and Policy Change in Ghana.* RAPnet: Case studies, http:www.gdnet.org/rapnet/research/studies/case_studies/Case_Study_25_Full.html.
Counts, Alex. *Small Loans, Big Dreams: How Nobel Prize Winner Muhammad Yunus and Microfinance Are Changing the World.* Hoboken, NJ: John Wiley & Sons, 2008.
Davis, Mike. *Planet of Slums.* New York: Verso, 2006.
de Botton, Alain. *The Architecture of Happiness.* London: Penguin, 2006.
Department of Housing/National Business Initiative: Housing Project Programming Guide. Pretoria, South Africa: Dept. of Housing, 1997.
de Soto, Hernando. *The Other Path: The Invisible Revolution in the Third World.* New York: Harper & Row, 1989.
Edstrom, Judith. *Indonesia's Kecamatan Development Project: Is It Replicable?* World Bank: Social Development Papers, Paper No. 39. Washington, DC: World Bank, 2002.
Festinger, Leon. *Deterrents and Reinforcement.* Stanford, CA: Stanford University Press, 1961.
Festinger, Leon, and Lawrence, Douglas H. *A Theory of Cognitive Dissonance.* Stanford, CA: Stanford University Press, 1957.
Fisher, Roger, Kopelman, Elizabeth, and Kupfer Schneider, Andrea. *Beyond Machiavelli: Tools for Coping with Conflict.* Cambridge, MA: Penguin, 1993.
Gladwell, Malcolm. *The Tipping Point: How Little Things Can Make a Big Difference.* London: Penguin, 2002.
———. *Blink: The Power of Thinking without Thinking.* London: Penguin, 2006.
Government of the Republic of South Africa. Constitution of South Africa Act. Pretoria, 1996.
———. National Building Regulations and Building Standards Act 103 of 1997. Pretoria, 1997.
———. National Health Act, Pretoria, 2004.
Government of the Republic of South Africa and Others v. Grootboom and Others. Constitutional Court of South Africa 2002 (1) SA 46 (CC).
Government of the Republic of Zambia. Housing (Statutory and Improvement Areas) Act. Lusaka: Government Printer, 1974.
Graham, Nick. "Informal Settlements Upgrading in Cape Town: Challenges, Constraints and Contradictions within Local Government." In *Informal Settlements—A Perpetual Challenge?* Ed. Marie Huchzermeyer and Aly Karam. Cape Town: UCT Press, 2006.
Greene, Solomon J. "Staged Cities: Mega-Events, Slum Clearance, and Global Capital." *Yale Human Rights and Development L.J.*, Vol. 6, 2003.
Hamdi, Nabeel. *Housing without Houses, Participation, Flexibility, Enablement.* London: Intermediate Technology Publications, 1995.
———. *Small Change: About the Art of Practice and the limits of Planning in Cities.* London: Earthscan, 2004.
Haro, Guyo, Doyo, Godana, and McPeak, John. "Linkages between Community, Environmental and Conflict Management: Experiences from Northern Kenya." *World Development*, Vol. 33, No. 2, 2005.
Hine, John, et al. Ghana Feeder Road Prioritisation, http://www.transport.links.org/transport_links/filearea/publications/1_789_pa3835.pdf.
Huchzermeyer, Marie. *Today's Tenement Cities: Large Scale Private Landlordism in Multi-Storey Districts of Nairobi Draft Research Report.* Johannesburg: University of the Witswatersrand, unpublished, 2006.

Huchzermeyer, Marie, et al. "Policy, Data and Civil Society: Reflections on South African Challenges through an International Review." In *Informal Settlements, a Perpetual Challenge?* ed Marie Huchzermeyer and Aly Karam. Cape Town: UCT Press, 2006.

Huntingdon, Samuel. *The Clash of Civilizations and the Remaking of World Order.* New York: Simon & Schuster, 1996.

Imperato, Ivo, and Ruster, Jeff. *Slum Upgrading and Participation: Lessons from Latin America.* Washington, DC: World Bank, 2003.

Jacobs, Jane. *The Economy of Cities.* New York: Vintage, 1970.

Jopson, Barney. "Clean Break." *Financial Times Weekend,* Life and Arts, November 29, 2008.

Kulipossa, Fidelx, and Manor, James. "The Direct Support to Schools Program in Mozambique." In *Aid That Works, Successful Development in Fragile States,* ed. James Manor. Washington, DC: World Bank, 2007.

Ledogar, Robert. *Community Participation, Collective Self-help and Community Development in the Lusaka Housing Project.* Lusaka, Zambia (mimeo), 1979.

Mabogunje, A. L., Hardoy, J. E., and Misra, R. P. *Shelter Provision in Developing Countries SCOPE Report 11.* Chichester, UK: John Wiley & Sons, 1978.

Manor, James, ed. *Aid that Works: Successful Development in Fragile States.* Washington, DC: World Bank.

Marris, Peter. *Family and Change in an African City: A Study of Rehousing in Lagos.* London: Routledge & Kegan Paul, 1961.

Martin, Brendan. *From Clientelism to Participation: The Story of 'Participatory Budgeting' in Porto Alegre.* unpublished paper, 1997.

Martin, R. J. *Gardens and Outdoor Living: Research Study No. 1.* Lusaka, Zambia: National Housing Authority, 1974.

———. *Self Help in Action.* Lusaka, Zambia: National Housing Authority, 1975.

Martin, Richard. "Toward a New Architecture." *RIBA Journal,* 10/76.

———. "Institutional Involvement in Squatter Settlements." *Architectural Design,* Vol. XLVI, April 1976.

———. "Experiences with Monitoring and Evaluation in Lusaka." In R. Skinner, J. Taylor, and E. Wegelin, *Shelter Upgrading for the Poor.* Manila: Island Publishing House, 1987.

Martin, Richard, and Ledogar, Robert. *A Squatter Settlement in Lusaka, Zambia "George Compound."* Report submitted to the United Nations Centre for Housing Building and Planning. Lusaka: Lusaka City Council, 1977.

Martin, Richard, and Mathema, Ashna. *Housing Finance in Morocco: Does It Work for the Poor?* Washington, DC: USAID, 2008.

Martin, Richard, Mathema Ashna, et al. *Mbabane Upgrading and Finance Project.* Washington, DC: Cities Alliance, 2007.

Martin, Richard, et al. *Development of Appropriate Housing Finance Products to Support Upgrading Activities.* Washington, DC: Cities Alliance, 2008.

Mathema, Ashna. *Mbabane Upgrading and Finance Project: Household Interviews.* For Mbabane City Council, funded by Cities Alliance/The World Bank, unpublished, 2005.

———. *Housing in Addis: Background Paper.* For the World Bank, funded by the Danish Trust Fund, unpublished, 2005.

———. *Eritrea: Housing and Urban Development Policy Study, Qualitative Study.* For Ministry of Public Works, Eritrea and UN-Habitat, unpublished, 2005.

———. *Nairobi's Informal Settlements: A View from the Inside.* Background Paper. Funded by the Norwegian Trust Fund/The World Bank, unpublished, 2005.

———. *Qualitative Study: Household Interviews, Accra, Ghana.* For African Union for Housing Finance, funded by Cities Alliance/The World Bank, unpublished, 2006.

————. *Slums and Sprawl: The Unintended Consequences of Well-Intended Regulation.* Washington, DC: World Bank, unpublished, 2007.

————. *Qualitative Study: Household Interviews, Dar es Salaam, Tanzania.* For African Union for Housing Finance, funded by Cities Alliance/The World Bank), unpublished, 2007.

————. *Planning and Practice: The Missing Link.* Paper prepared for the International Conference on Homelessness: a Global Perspective, Delhi, unpublished, 2007.

Mathema, Ashna, and Nayana, Mawilmada. *Decentralization and Housing Delivery: Lessons from the Case of San Fernando, La Union, Philippines.* Unpublished thesis for Master of City Planning program at MIT, 2000.

National Building Standards, 1996, section 13. Government of Ghana, L.I. 1630.

National Housing Authority. *Criteria for the Design of Low Cost and Very Low Cost Houses.* Lusaka: National Housing Authority, 1972.

Neuwirth, Robert. *Shadow Cities, A Billion Squatters, a New Urban World.* New York: Routledge, 2005.

Njonjo, Apollo. *Study on Community Managed Water Supplies,* Draft. Nairobi: Report on Case Studies for the World Bank Regional Water and Sanitation Group— East Africa, UNDP/WB Water and Sanitation Programme, 1994, quoted in Deepa Narayan, *Designing Community Based Development.* Rome: SD Dimensions, UN Food and Agricultural Organisation, Sustainable Development Department,1997.

Olken, Benjamin A. *Monitoring Corruption: Evidence from a Field Experiment in Indonesia.* NBER Working Paper 11753. National Bureau of Economic Research. Cambridge, MA: National Bureau of Economic Research, 2005.

PADCO. *Metro Manila Urban Services for the Poor Project.* Washington, DC: PADCO, 2003.

————. *Study on Upgrading in the Peri-Urban Areas of Swaziland.* Prepared for the World Bank and the Ministry of Urban Development and Housing. Washington, DC: PADCO, 2003.

Panos Pastoralist Communication Initiative. 2003, http://www.comminit.com/strategicthinking/stfaocommnrm/sld-1700.html.

Pasteur, David. *The Management of Squatter Upgrading: A Case Study of Organisation, Procedures and Participation.* Farnborough, UK: Saxon House, 1979.

Patel, Sheela, and Arputham, Jockim. "An Offer of Partnership or a Promise of Conflict in Dharavi, Mumbai?" *Environment and Urbanization,* Vol. 19, No. 2, October 2007.

Perlman, Janice E. *The Myth of Marginality: Urban Poverty and Politics in Rio de Janeiro.* Berkeley: University of California Press, 1976.

————. *Favela: Four Decades of Living on the Edge in Rio de Janeiro.* New York, Oxford University Press. Forthcoming.

Polar, Paul. *Out of Poverty: What Works When Traditional Approaches Fail.* San Francisco: Berrett-Koehler, 2008.

Potter, Alana, and Skinner, Kate. *Lessons Learned for Improving Communication at the Local Level.* Paper presented at the 11th ITN Africa Conference, Harare, Zimbabwe, 1999.

Prahalad, C. K. *Fortune at the Bottom of the Pyramid: Eradicating Poverty through Profits.* Upper Saddle River, NJ: Wharton School Publishing, 2006.

Putnam, Robert. *Bowling Alone: The Collapse and Revival of American Community.* New York: Simon & Schuster, 2000.

Riis, Jacob A. *How the Other Half Lives.* New York: Dover, 1971.

Robinson, Mark. "Community-Driven Development in Conflict and Postconflict Conditions." In *Successful Development in Fragile States,* ed. James Manor. Washington, DC: World Bank, 2007.

Sachs, Jeffrey. *The End of Poverty: How We Can Make It Happen in Our Lifetime.* London: Penguin, 2005.

Samaj, Bharat Sevak. *The Slums of Old Delhi*. Delhi: Arma Ram and Sons, 1958.
Savchuk, Katia, Echanove, Matias, and Srivastava Rahul. *Intro: Lakhs of Residents, Billions of Dollars*, http://www.dharavi.org.
Schlyter, Ann, and Schlyter, Thomas. *George—the Development of a Squatter Settlement in Lusaka, Zambia*. Lund, Sweden: Swedish Council for Building Research, 1979.
Schwartz, Norman, and Derruyterre, Anne. *Community Consultation, Sustainable Development and the Inter-American Development Bank*. Washington, DC: Inter-American Development Bank, 1996.
Sen, Amartya. *Development as Freedom*. Oxford: Oxford University Press, 1999.
Shack Slum Dwellers Association: Indian Alliance, Seeing is Believing. Delhi: Mahila Milan/National Slum Dwellers' Federation N.S.D.F/Society for the Promotion of Area Resource Centre (SPARC), 2004.
Shah, Parmesh, "Participation in Poverty Reduction." *North-South Institute, e-Views*, Issue 7, 2000.
Shanahan, Peter. *Report on Community Development Training Programme and Proposals for Future Programmes*. Lusaka, Zambia: Housing Project Unit, Lusaka City Council, mimeo, 1976.
Sinha, Rama. *Three Jeers for Bureaucracy*. Delhi: Orient Longman, 1978.
Smit, Warren. "Understanding the Complexities of Informal Settlement: Insights from Cape Town." In *Informal Settlements—A Perpetual Challenge?* Ed. Marie Huchzermeyer, and Aly Karam, Cape Town: UCT Press, 2006.
Smith, Daniel Jordan. *A Culture of Corruption: Everyday Deception and Popular Discontent in Nigeria*. Princeton, NJ: Princeton University Press, 2006.
Suroor, Hasan. "Slums Better Equipped for Challenges: Charles." *The Hindu*, February 7, 2009.
Turner, John F. C. *Housing by People*. London: Marion Boyars, 1976.
Turner, John F. C., and Fichter, Robert, eds. *Freedom to Build*. New York: Collier Macmillan, 1972.
UN Habitat. *Slums of the World: The Face of Urban Poverty in the New Millennium?* Nairobi: United Nations Human Settlements Programme, 2003.
USAID/G/HCD/HETS. Monitoring Training for Results. Washington, DC: AMEX International, Inc., Creative Associates, Inc., 1996.
Vance, Irene. "The Community as Evaluators: Experience with Community Participation in Self-Build Projects in Managua, Nicaragua." In R. Skinner, J. Taylor, and E. Wegelin, *Shelter Upgrading for the Urban Poor*. Manila: Island Publishing House, 1987.
Village Education Resource Center. *Shifting Millions from Open Defecation to Hygienic Latrines, Process Documentation of 100% Sanitation Approach*. Dacca: VERC, 2002.
Yahya, S., et al. *Double Standards, Single Purpose*. London, Intermediate Technology Development Group, 2001.
Young, Michael, and Willmott, Peter. *Family and Kinship in East London*. London: Penguin, 1957.
World Bank. *Project Appraisal Document on a Proposed Credit to the Government of Uganda for a Northern Uganda Social Action Fund*, Report 23885. Washington, DC: World Bank, 2002.
———. *Lagos Metropolitan Development and Governance Project Appraisal Report*. Washington, DC: World Bank, 2006.
———. *World Development Report, 2009, Reshaping Economic Geography*. Washington, DC: World Bank, 2009.
Zimbardo, Philip. *The Lucifer Effect: How Good People Turn Evil*. London: Rider, 2007.

Index

For Product Safety Concerns and Information please contact our EU
representative GPSR@taylorandfrancis.com
Taylor & Francis Verlag GmbH, Kaufingerstraße 24, 80331 München, Germany